International Series in Operations Research & Management Science

Volume 108

For other titles published in this series, go to
http://www.springer.com/series/6161

Dirk Beyer · Feng Cheng · Suresh P. Sethi · Michael Taksar

Markovian Demand Inventory Models

Springer

Dirk Beyer
M-Factor
1400 Fashion Island Blvd.
Suite 602
San Mateo CA 94404
USA
dirk.beyer@m-factor.com

Feng Cheng
Federal Aviation Administration
Office of Performance Analysis and Strategy
800 Independence Ave. S.W.
Washington, D.C. 20591
USA
feng.cheng@faa.gov

Suresh P. Sethi
School of Management, M/S SM30
The University of Texas at Dallas
800 W. Campbell Road, SM 30
Richardson, TX 75080-3021
USA
sethi@utdallas.edu

Michael Taksar
Department of Mathematics
University of Missouri
Rollins Road
Columbia MO 65211
USA
taksar@math.missouri.edu

ISSN 0884-8289
ISBN 978-1-4614-2466-6 e-ISBN 978-0-387-71604-6
DOI 10.1007/978-0-387-71604-6
Springer New York Dordrecht Heidelberg London

Printed on acid-free paper

Springer is part of Springer Science+Business Media (www.springer.com)

Table of Contents

Part V CONCLUSIONS AND OPEN RESEARCH PROBLEMS

Part VI APPENDICES

List of Figures

List of Tables

Preface

Inventory management is among the most important topics in operations management and operations research. It is perhaps also the earliest, as evidenced by the economic order quantity (EOQ) formula, which dates back to Ford W. Harris (1913). Inventory management is concerned with matching supply with demand. The problem is to find the amount to be produced or purchased in order to maximize the total expected profit or minimize the total expected cost.

In the two decades following Harris, several variations of the formula appeared, mostly in trade journals written by and for inventory managers. These were first collected in a full-length book on inventory management by Raymond (1931). Erlenkotter (1989,1990) has provided a fascinating account of this early history of inventory management.

Another classic formula in the inventory literature is the newsvendor formula. It can be attributed to Edgeworth (1888), although it was not until the work of Arrow, Harris and Marschak (1951) and Dvoretzky, Kiefer and Wolfowitz (1953) that a systematic study of inventory models, incorporating uncertainty as well as dynamics, began. These early papers were followed by the well-known treatise by Arrow, Karlin and Scarf (1958) as well as the seminal work of Scarf (1960). These works established the optimality of base-stock and (s, S) policies, the most well-known policies in dynamic stochastic inventory modeling.

An important assumption in this vast literature has been that the demands in different periods are independent and identically distributed (i.i.d.). In real life, demands may depend on environmental considerations or the states of the world, such as the weather, the state of the economy, etc. Moreover, these states of the world are represented by stochastic processes – exogenous or controlled.

In this book, we are concerned with inventory models where these world states are modeled by Markov processes. We then show that we obtain the optimality of (s, S)-type policies, or base-stock policies (i.e., $s = S$) when there are no fixed ordering costs with the provision that the policy parameters s and S depend on the current state of the Markov process representing the environment. Models allowing backorders when the entire demand cannot be filled from the available inventory, as well as those when the current demand is lost, are considered. As for the cost criterion, we treat both the minimization of the expected total discounted cost and the long-run average cost. The average cost criterion is mathematically more difficult than the discounted cost criterion. Finally, we generalize the usual assumptions on holding and shortage costs and on demands that are made in the literature.

Our research on Markovian demand inventory models was carried out over a period of ten years, beginning in the early nineties. The research was supported by the National Science and Engineering Research Council of Canada, the Manufacturing Research Corporation of Ontario (a provincial Center of Excellence), the Laboratory for Manufacturing Research and The Canadian Centre for Marketing Information Technologies at the University of Toronto, and the University of Texas at Dallas. The first and second authors gratefully acknowledge the support from Alexander v. Humboldt Foundation and IBM Corporation, respectively.

This research has appeared in several journals, and it is now the subject of this book.

The models and the results presented in this book could be used in the analysis of inventory models with forecast updates. This topic is covered in the book by Sethi *et al.* (2005a).

Mathematical tools employed in this book involve dynamic programming and stochastic processes. This book is written for students, researchers, and practitioners in the areas of operations management and industrial engineering. It can also be used by those working in the areas of operations research and applied mathematics.

We are grateful to Alain Bensoussan, Ganesh Janakiraman, Ernst Presman, Selda Taskin, Yusen Xia, Xiangtong Qi, Hanqin Zhang, Jun Zhang, Qing Zhang, Arnab Bisi, and Maqbool Dada for their careful reading of parts of the manuscript. In addition, we express our appreciation to Barbara Gordon for her assistance in the preparation of the manuscript.

San Mateo, CA, USA, September, 2009 — Dirk Beyer
Washington D.C., USA, September, 2009 — Feng Cheng
Richardson, TX, USA, September, 2009 — Suresh P. Sethi
Columbia, MO, USA, September, 2009 — Michael Taksar

To my wife Michelle and sons Robert and Harry — *Dirk Beyer*

To my wife Yanyu and son Eric — *Feng Cheng*

To my wife Andrea
and daughters Chantal and Anjuli — *Suresh P. Sethi*

To my wife Tanya and son Serge — *Michael Taksar*

Notation

This book is divided into six parts containing ten chapters 1,2,...,10 and three appendices A, B, and C. Each of the ten chapters in this book is divided into sections. In any given chapter, say Chapter 8, sections are numbered consecutively as 8.1, 8.2, 8.3, and so on. Subsections within a section, say Section 8.4, are numbered 8.4.1, 8.4.2, ... Sub-subsections within a subsection, say Subsection 8.4.2, are numbered consecutively as 8.4.2.1, 8.4.2.2, ... Mathematical expressions within a chapter such as equations, inequalities, and conditions, are numbered consecutively as (8.1), (8.2), (8.3), ... Theorems, lemmas, corollaries, definitions, remarks, examples, figures and tables within a chapter are also numbered consecutively such as Theorem 8.1, Theorem 8.2, Theorem 8.3, ...

Each appendix, say Appendix B, has sections numbered consecutively as B.1, B.2, B.3, ... Subsections within a section, say Section B.2, are numbered consecutively as B.2.1, B.2.2, B.2.3, ... Mathematical expressions within an appendix are also numbered consecutively such as (B.1), (B.2), (B.3), ... Unlike the numbering in the chapters, theorems, lemmas, corollaries, definitions, and remarks within a section, say Section B.5, are numbered consecutively such as Theorem B.5.1, Theorem B.5.2, ...

The terms "surplus", "inventory/shortage", and "inventory/backlog" are used interchangeably. The terms "control", "policy", and "decision" are used interchangeably.

The order of the appearance of the notation used in this book is symbols, numerals, capital letters in alphabetical order, lower case letters in alphabetical order, Greek letters in alphabetical order regardless of their case, and abbreviations.

$\square$	indicates end of a proof, example, definition, or remark
$:=$	is defined to be equal to
$\Longrightarrow$	implies
$\langle 0, N\rangle$	$= \{0, 1, 2, \ldots, N\}$, horizon of the inventory problem
$\mathbf{0}$	$(0, \ldots, 0)$, the policy of ordering nothing in every period

$\mathbb{1}_A$	indicator function of subset $A \subset \Omega$, i.e., $\mathbb{1}_A(\omega) = 1$ if $\omega \in A$ and $\mathbb{1}_A(\omega) = 0$, otherwise. Equivalently, A can refer to a condition and $\mathbb{1}_A = 1$ if condition A holds and $\mathbb{1}_A = 0$, otherwise
$\mathbb{B}(F)$	set of real Borel functions defined on a set F
$\mathbb{B}_0$	class of all continuous functions from $\mathbb{I} \times \mathbb{R}$ into $\mathbb{R}^+$ and pointwise limits of sequences of these functions
$\mathbb{B}_1$	subspace of functions in $\mathbb{B}_0$ that are lower semicontinuous and of linear growth
$\mathbb{B}_2$	subspace of functions in $\mathbb{B}_1$ that are uniformly continuous in x
$\mathbb{B}^\gamma$	Banach space of Borel functions $b : \mathbb{I} \times \mathbb{R} \to \mathbb{R}$ with the norm $\| b \|_\gamma = \max_i \sup_x \|b(i,x)\|/(1+\|x\|^\gamma) < \infty$
$\mathbb{C}([0,T])$	space of all continuous functions on $[0,T]$
$\mathbb{C}_1$	subspace of functions in $\mathbb{B}_1$ that are uniformly continuous
$\mathbb{D}$	convex subset $\mathbb{D} \subset \mathbb{R}$
$\mathsf{E}\xi$	expectation of a random variable ξ
$\mathsf{E}(\xi; A)$	$\mathsf{E}(\xi \mathbb{1}_A)$ for any random variable ξ and any set $A \subset \Omega$
$\mathsf{E}\{\xi_k \| i_k = i\}$	conditional expectation of ξ_k when $i_k = i$
$F_1 \cap F_2$	intersection of sets F_1 and F_2
$F_1 \cup F_2$	union of sets F_1 and F_2
$\mathbb{I}$	$\langle 1, L\rangle = \{1, 2, \ldots, L\}$, finite collection of possible demand states
$J_n(i, x; U)$	objective function representing the cost to go under decision U from period n on, with the initial conditions i and x at n
$J^\alpha(i, x; U)$	objective function for the infinite horizon problem with dependence on the discount factor α made explicit
K_n^i	fixed order cost in period k if $i_k = i$
$\mathbb{L}^\gamma$	subspace of lower semicontinuous functions in $\mathbb{B}^\gamma$
$\mathbb{L}^{\gamma^-}$	class of all lower semicontinuous functions which are of polynomial growth with power γ or less on $(-\infty, 0]$
$\mathbb{R}$	$= (-\infty, \infty)$, the real line
$\mathbb{R}^+$	$= [0, \infty)$
$\mathbb{R}^n$	n-dimensional Euclidean space
$S, S_i, S_{n,i}, S^*$	order-up-to level; i refers to state i and n refers to period n,
S_i^α	order-up-to level when the state is i with dependence on the discount factor α made explicit
U, U^α	$\mathrm{U} = (u_0, u_1, \ldots)$, $u_i \geq 0$, $i = 0, 1, \ldots,$ a history-dependent or nonanticipative decision

$\mathbb{Z}$	$= \{0, 1, 2, \dots\}$
a^+	$= \max\{a, 0\}$ for a real number a
a^-	$= \max\{-a, 0\}$ for a real number a
$[a, b]$	set $\{x \mid a \le x \le b\}$
$[a, b)$	set $\{x \mid a \le x < b\}$, similarly $(a, b]$ and (a, b)
$\| b \|, \| b \|_\gamma$	norm of b in $\mathbb{B}_1$ and $\mathbb{B}^\gamma$, respectively
$\bar{f}(i)$	$= \sum_{j=1}^{L} p_{ij} f(j) \equiv \mathsf{E} f(i_1 \mid i_0 = i)$, if $f(\cdot)$ is a function on $\mathbb{I}$
$f_k(i, x)$	surplus cost when $i_k = i$ and $x_k = x$
g', g'^+, g'^-	derivative, right derivative, and left derivative of function g, respectively
i_k	demand state in period k
$\{i_k\}$	a Markov chain with an $(L \times L)$-transition matrix $P = \{p_{ij}\}$
$\{l\}$	sequence $(l = 1, 2, \dots)$ or a subsequence of it
$\langle n, N \rangle$	$\{n, n+1, \dots, N\}$
$\{p_{ij}\} = P$	transition matrix
$q_k(i, x)$	shortage cost in period k when $i_k = i$ and $x_k = x$
$s, s_1, s_{n,i}, s^*$	ordering level; i refers to state i and n to period n,
s_i^α	ordering level when the state is i with dependence on the discount factor α made explicit
u_k, u_k^α, u_k^*	nonnegative order quantity in period k
$v, v_n, v_{n,k}, \dots$	value function
v^α	value function for the infinite horizon with dependence on the discount factor α made explicit
w, w^*	potential function
w^α	differential discounted value function
$\boldsymbol{x}, \boldsymbol{z}, \dots$	all English boldface letters stand for vectors
$\|x\|$	absolute value of x
$\| \boldsymbol{x} \|$	$= \sqrt{x_1^2 + \cdots + x_n^2}$ for a vector $\boldsymbol{x} = (x_1, \dots, x_n)$
x_k	surplus (inventory/backlog) level at the *beginning* of period k
$x_n \to x$	x_n converges to x
$x_n \downarrow x$	x_n decreases to x
$x_n \uparrow x$	x_n increases to x
$\alpha > 0$	discount factor
α_p	Type 1 service level; $0 < \alpha_p \le 1$
λ, λ^*	average cost
ξ_k	demand in period k, $\xi_k \ge 0$, ξ_k dependent on i_k
$\sigma\{\mathbf{k}(s) : s \le t\}$	σ-algebra generated by the process $\mathbf{k}(\cdot)$

τ_y	$:= \min\{n : \sum_{k=0}^{n} \xi_k \geq y\}$ the first time the cumulative demand exceeds y
$\varphi_{i,k}(\cdot), \varphi_i(\cdot)$	conditional density function of demand ξ_k when $i_k = i$
$\Phi_{i,k}(\cdot), \Phi_i(\cdot)$	conditional distribution functions corresponding to $\varphi_{i,k}$ and φ_i, respectively
$(\Omega, \mathcal{F}, \mathsf{P})$	probability space

a.s.	almost surely
EOQ	economic order quantity
i.i.d.	independent and identically distributed
inf	infimum
$\lim_{x\downarrow 0}$	limit when x approaches 0 from right
$\lim_{x\uparrow 0}$	limit when x approaches 0 from left
lim inf	limit inferior
lim sup	limit superior
LHS	left-hand side
l.s.c.	lower semicontinuous
max	maximum
MDP	Markov decision process
min	minimum
PF_2	Pólya frequency of order 2
RHS	right-hand side
sup	supremum
u.s.c.	upper semicontinuous

I

INTRODUCTION

Chapter 1

INTRODUCTION

Inventory management is one of the most important tasks in business. A business faces inventory problems in its most basic activities. Inventory is held by the selling party to meet the demand made by the buying party. The complexity of inventory problems varies significantly, depending on the situation. While some of the simple inventory problems may be dealt with by common sense, some other inventory problems that arise in complex business processes require sophisticated mathematical tools and advanced computing power to get a reasonably good solution.

The fundamental problem in inventory management can be described by the following two questions: (1) when should an order be placed? and (2) how much should be ordered? There are two basic trade-offs in an inventory problem. One is the trade-off between setup costs and inventory holding costs. By placing orders frequently, the size of each order can be made relatively small. Therefore, the holding costs can be reduced. However, the total setup costs will go up. Conversely, less frequent orders will save on setup costs but incur higher holding costs. The other trade-off is between holding costs and stockout costs. Holding more inventory reduces the likelihood of stockouts, and vice versa. These trade-offs give rise to an optimization problem of finding the optimal ordering policy that minimizes the overall cost.

1.1. Characteristics of Inventory Systems

The primary purpose of inventory control is to manage inventory to effectively meet demand. The effectiveness can be measured in many different ways. The most commonly used methods measure how much the total profit and how little the total cost are. The majority of the

D. Beyer et al., *Markovian Demand Inventory Models*, International Series in Operations Research & Management Science 108,
DOI 10.1007/978-0-387-71604-6_1,

operations research literature on inventory management has used the criterion of cost minimization. In many cases, this criterion is equivalent to that of profit maximization.

There are many factors that should be taken into consideration when solving an inventory problem. These factors are usually the characteristics of, or the assumptions made about, the particular inventory system under consideration. Among them, the most important ones are listed below.

Cost structure. One of the most important prerequisites for solving an inventory problem is an appropriate cost structure. A typical cost structure incorporates the following four types of costs.

- *Purchase or production cost.* This is the cost of buying or producing items. The total purchase/production cost is usually expressed as cost per unit multiplied by the quantity procured or produced. Sometimes a quantity discount applies if a large number of units are purchased at one time.
- *Fixed ordering (or setup) cost.* The fixed ordering cost is associated with ordering a batch of items. The ordering cost does not depend on the number of items in the batch. It includes costs of setting up the machine, costs of issuing the purchase order, transportation costs, receiving costs, etc.
- *Holding (or carrying) cost.* The holding or carrying cost is associated with keeping items in inventory for a period of time. This cost is typically charged as a percentage of dollar value per unit time. It usually consists of the cost of capital, the cost of storage, the costs of obsolescence and deterioration, the costs of breakage and spoilage, etc.
- *Stockout cost.* Stockout cost reflects the economic consequences of unsatisfied demand. In cases when unsatisfied demands are backlogged, there are costs for handling backorders as well as costs associated with loss of customer goodwill on account of negative effects of backlogs on future customer demands. If the unsatisfied demand is lost, i.e., there is no backlogging, then the stockout cost will also include the cost of the foregone profit.

Demand. Over time, demand may be constant or variable. Demand may be known in advance or may be random. Its randomness may depend on some exogenous factors such as the state of the economy, the weather condition, etc. Another important factor often ignored in the inventory literature is that demand can also be influenced directly or indirectly by the decision maker's choice. For example, a promotion decision can have a positive effect on demand.

Leadtime. The leadtime is defined as the amount of time required to deliver an order after the order is placed. The leadtime can be constant (including zero) or random.

Review time. There are two types of review methods. One is called continuous review, where the inventory levels are known at all times. The other is called periodic review, where inventory levels are known only at discrete points in time.

Excess demand. Excess demand occurs when demand cannot be filled fully from the existing inventory. The two common assumptions are that excess demand is either backordered or lost.

Deteriorating inventory. In many cases, the inventory deteriorates over time which affects its utility. For example, food items have limited shelf lives.

Constraints. There are various constraints involved in inventory problems. The typical constraints are supplier constraints – restrictions on order quantities; marketing constraints – minimum customer service levels; and internal constraints – storage space limitations, limited budgets, etc.

1.2. Brief Historical Overview of Inventory Theory

Inventory control has been a main topic in the operations management area for over half a century. Many advanced mathematical methods can be applied to solve inventory control problems. This makes it an interesting topic for researchers from a variety of academic disciplines. The basic question in any inventory problem is to determine the timing and the size of a replenishment order. While the fundamental questions in inventory problems remain the same, the focus of one particular inventory problem may be quite different from another.

It is virtually impossible to summarize the enormous literature on inventory models in this short section. Even with the limitation of focusing on studies aimed at finding optimal policies for dynamic inventory models, it is still a tremendous task to cover all related issues. Different features of inventory systems also lead to different treatments of the resulting problems. Silver (1981) has illustrated such a diversity of inventory problems by introducing a classification scheme consisting of the following characteristics of an inventory system.

Single vs. Multiple Items

Deterministic vs. Probabilistic Demand

Single Period vs. Multiperiod

Stationary vs. Time-Varying Parameters

Nature of the Supply Process

Procurement Cost Structure

Backorders vs. Lost Sales

Shelf Life Considerations

Single vs. Multiple Stocking Points

Among various factors that affect the model characterization, the nature of the demand process is usually the most important one. Assumptions about other parameters of an inventory model also have important implications on the model development. However, these assumptions are more or less standard and well-accepted in practice.

In this book, we will consider finite and infinite horizon inventory models involving a single product, a single stocking point, and a stochastic demand. We will assume that supply is unconstrained and the product has an infinite shelf life. We will allow the system to be with either stationary or time-varying parameters. Both backorders and lost sales models will be treated. We will consider zero leadtime. However, the backlogging models with fixed nonzero leadtimes can be transformed to ones with no leadtime.

The main distinguishing characteristic of the inventory models studied in this book is that the demand will be dependent on a Markov process. The Markov process may be exogenous (i.e., outside our control) or it may be influenced by our inventory decisions.

In what follows, we will briefly summarize the relevant literature based on the assumptions made about the demand process in a single product setting.

1.2.1 Deterministic Demand Models

The most celebrated model in this category is the economic order quantity (EOQ) model, which is the simplest and the most fundamental of all inventory models. It describes the important trade-off between the fixed ordering cost and the holding cost, and is the basis for the analysis of many more complex systems.

The basic EOQ model is based on the following assumptions.

(i) The demand rate is a constant ξ units per unit time.

(ii) Shortages are not permitted.

(iii) The order leadtime is zero.

(iv) The order costs consist of a setup cost K for each ordered batch and a fixed unit cost c for each unit ordered.

(v) The holding cost is h per unit held per unit time.

The famous EOQ formula is given by

$$Q^* = \sqrt{\frac{2K\xi}{h}}.$$

Because of its form, it is also called the *square-root formula; (see also EOQ formula).* The formula is due to Harris (1913); (see Erlenkotter (1989) for a history of the development of this formula).

Some important variations of the EOQ model include planned back-orders/lost sales, quantity discounts, a known shelf life, a known replenishment leadtime, or constraints on the order size. A fairly comprehensive discussion on these and related topics can be found in Hadley and Whitin (1963).

Another important work in deterministic demand models is the so-called dynamic lot size problem, where the demand level changes from period to period but in a known fashion. Wagner and Whitin (1958) provide a recursive algorithm for computing the optimal policy. Some interesting properties of the optimal policy are identified by them. The Wagner-Whitin model was republished in a 2004 issue of *Management Science* with a commentary by Wagner (2004).

1.2.2 Stochastic Demand Models

The newsvendor formula for the single period inventory problem is one of the most important results in the stochastic inventory theory; (see Edgeworth (1888), Arrow *et al.*(1951)). Many extensions of the classical newsvendor model have been studied in the literature. These allow for random yields, pricing policies, free distributions, etc. A comprehensive review of the literature on the newsvendor models can be found in Khouja (1999), and references therein.

The base-stock policy has been extensively studied in the literature for single-product periodic review inventory problems with stochastic demand and no fixed setup cost. The early academic discussions on the base-stock policy have focused on proving its optimality under the situation that the demands are independent. Examples can be found in

Gaver (1959,1961), Karlin (1958c), Karlin and Scarf (1958), etc. Bellman *et al.* (1955) and Karlin (1960) are the classical papers for the stationary and nonstationary demand cases, respectively. Optimality of a base-stock policy has been established in these situations.

For single-item models with probabilistic demand and a fixed ordering cost, it has been shown that an (s, S) policy is optimal under a variety of conditions. The policy is to order up to the level S when the inventory level is below s and to not order otherwise. In the case of no fixed ordering cost, the optimal policy becomes a base-stock policy, which is a special case of the (s, S) policy when $s = S$.

Classical papers on the optimality of (s, S) policies in dynamic inventory models with stochastic demands and fixed setup costs are those of Arrow *et al.*(1951), Dvoretzky *et al.* (1953), Karlin (1958c), Scarf (1960), and Veinott (1966). Scarf develops the concept of K-convexity, and uses it to show that if the ordering cost is linear with a fixed setup cost K and if the inventory/backlog cost function is convex, then the optimal ordering policy in any given period for a finite horizon model is characterized by two critical numbers s_n and S_n, with $s_n \leq S_n$, such that if the inventory level x_n at time n is less than s_n, then order $S_n - x_n$; if $x_n \geq s_n$, then do not order. That a stationary (s, S) policy is optimal for the stationary infinite horizon problem is proved by Iglehart (1963a). Optimality of an (s, S) policy in the lost sales case was proved by Shreve (1976). Proof in the lost sales case does not extend to the case when the leadtime is nonzero; (see, e.g., Zipkin (2008b)).

Veinott (1966) presents a model similar to Scarf's but with a different set of conditions, which neither imply nor are implied by the conditions used by Scarf (1960). Instead of using the concept of K-convexity, he proves the optimality of an (s, S) policy by showing that the negative of the expected cost is a unimodal function of the initial inventory level. His model also provides a unified approach for handling both backlogging and lost sales assumptions, as well as the case of perishable products.

1.2.3 Markovian Demand Models

Most classical inventory models assume demand in each period to be a random variable independent of environmental factors other than time. However, many randomly changing environmental factors, such as fluctuating economic conditions and uncertain market conditions in different stages of a product life cycle, can have a major effect on demand. For such situations, the Markov chain approach provides a natural and flexible alternative for modeling the demand process. In such an approach, environmental factors are represented by the demand state or *the state-of-the-world* of a Markov process, and demand in a period is a random variable with a distribution function dependent on the demand state in that period. Furthermore, the demand state can also affect other parameters of the inventory system such as the cost functions.

The effect of a randomly changing environment in inventory models with fixed costs received only limited attention in the early literature. It is conceivable that when the demands in different periods are dependent, the structure of the optimal policy does not change. The main difference is that the parameters of an optimal policy will depend on the demand state in the previous period.

Iglehart and Karlin (1962) consider the problem with dependent demand and no setup cost. They consider an inventory model with the demand process governed by a discrete-time Markov chain. In each period, the current value of the state of the chain decides the demand density for that period. They prove that the optimal policy is a state-dependent base-stock policy.

Karlin and Fabens (1960) introduce a Markovian demand model in which demand depends on the state of an underlying Markov chain. They indicate that given the Markovian demand structure in their model, it appears reasonable, in the presence of a fixed ordering cost, to postulate an inventory policy of (s, S)-type with a different set of critical numbers for each demand state. However, due to the complexity of the analysis, Karlin and Fabens concentrated on optimizing over the class of ordering policies, with each policy characterized by a single pair of critical numbers s and S, irrespective of the demand state.

Song and Zipkin (1993) present a continuous-time formulation with a Markov-modulated Poisson demand and with linear costs of inventory and backlogging. They show the optimality of a state-dependent (s, S) policy for the case when there is a fixed ordering cost. An algorithm for computing optimal policies is also developed using a modified value iteration approach. Sethi and Cheng (1997) provide a discrete-time version of the problem with general demand distributions. They show the optimality of a state-dependent (s, S) policy when the demand is modeled as a Markov-modulated process.

Cheng and Sethi (1999b) extend the results to the case of Markov-modulated demand with the lost sales assumption for unfilled demand. The optimality of a state-dependent (s, S) policy is proved for this case only under the condition of zero supply leadtime.

Another notable development in inventory models with Markovian demand is due to Chen and Song (2001). Their paper considers a multi-stage serial inventory system with a Markov-modulated demand. Random demand arises at Stage 1, Stage 1 orders from Stage 2, etc., and

Stage N orders from an outside supplier with unlimited stock. The demand distribution in each period is determined by the current state of an exogenous Markov chain. Excess demand is backlogged. Linear holding costs are incurred at every stage, and linear backorder costs are incurred at Stage 1. The ordering costs are also linear. The objective is to minimize the long-run average cost in the system. The paper shows that the optimal policy is an echelon base-stock policy with state-dependent order-up-to levels. An efficient algorithm is also provided for determining the optimal base-stock levels. The results can be extended to serial systems in which there is a fixed ordering cost at stage N and to assembly systems with linear ordering costs.

1.2.4 Models with Controllable Demands

When marketing tools such as promotions are used to stimulate consumer demand, the coordination between marketing and inventory management decisions becomes important. While it is clear that an integrated decision-making system should be adopted in order to maximize total benefits and avoid potential conflicts, little theoretical work can be found in the literature to guide such integrated decision making. Most works in the inventory theory do not deal with promotion decisions explicitly. Usually, promotion decisions are taken as given so that the effect of promotions can be determined in advance of inventory replenishment decisions. Thus, the inventory/promotion problem reduces to a standard inventory control problem with demand unaffected by the actions of the decision maker. However, there have been a few papers that address the issue of joint inventory/promotion decision making. Most of these focus on inventory control problems in conjunction with price discount decisions.

Used frequently in the literature, a simplified approach for modeling the demand uncertainty is to assume that the random components of demand are either additive or multiplicative; (see Young (1978)). Based on this assumption, Karlin and Carr (1962), Mills (1962), and Zabel (1970) obtained similar results. They showed that the optimal price is greater than the riskless price in the multiplicative case and is less than the riskless price in the additive case. Furthermore, Thowsen (1975) showed that if the demand uncertainty is additive, the probability density function of the additive uncertainty is PF_2, and if the expected demand function is linear, then the optimal policy is a base-stock list price policy. If the initial inventory level is below the base-stock level, then that stock level is replenished and the list price is charged; if the initial inventory level is above the base-stock level, then nothing is ordered and a price discount is offered.

A dynamic inventory model with partially controlled additive demand is analyzed by Balcer (1983), who presents a joint inventory control and advertising strategy and shows it to be optimal under certain restrictions. However, in the Balcer model, the demand/advertising relationship is deterministic because only the deterministic component of demand is influenced by advertising decisions.

Cheng and Sethi (1999a) extend the result of Markov-modulated demand models by introducing promotional decisions by which demand is affected. For the case of no fixed ordering cost, they show that a state-dependent base-stock policy is optimal. Chen and Simchi-Levi (2004) analyze a single-product, periodic-review model in which pricing and production/inventory decisions are made simultaneously. They show that when the demand model is additive, the profit-to-go functions are K-concave; hence, an (s, S, p) policy is optimal. In such a policy, the inventory is managed using an (s, S) policy, and price is determined based on the inventory position at the beginning of each period.

1.2.5 Other Extensions

Beside different treatments for modeling the demand process, extensions intended to relax other restrictive assumptions on inventory models or to incorporate certain special features into inventory models have also been studied by researchers. A great deal of effort has been made to adopt a more general cost function rather than the linear cost with a setup cost. An interesting case of such a general cost function is that of concave increasing cost. Porteus (1971) shows a generalized (s, S)-type policy to be optimal when the ordering cost is concave and increasing, given an additional assumption that the demand density function is a Pólya frequency function.

Other extensions for single-product inventory models include incorporating more realistic features of the physical inventory systems, such as capacity constraints, service level constraints, and supply constraints. Veinott (1965) considers a model where orders must be placed in multiples of some fixed batch size Q. It is shown that an optimal policy is characterized by Q and a number k as follows. If the initial inventory position is less than k, one orders the smallest multiple of Q, which brings the inventory position to at least k; otherwise, no order is placed.

Among the recent research trends in the analysis of dynamic inventory systems, one particular issue is to model the effect of more flexible procurement options such as multiple delivery modes in connection with the demand forecast updates available over time. Sethi *et al.* (2003) present an inventory model with fixed costs, forecast updates, and two delivery modes, with a forecast-update-dependent (s, S)-type policy shown to be optimal. This and related models are also the subject of the book by Sethi *et al.* (2005a).

Recently, Bensoussan *et al.* (2007a,2008a) have considered Markovian demand models with partially observed demands. In these models, called the censored newsvendor models, only sales are observed when the demand exceeds the inventory level. In other models, the information regarding the inventory is incomplete; (see, e.g., Bensoussan *et al.* (2005a,2007b,2008b)) .

There are several other related works extending the classical inventory models in various ways. We do not intend to provide a complete list of all relevant work in this fairly broad and active research area. Instead, we have chosen to survey only those that are related to the topic of this book either from a historical point of view or because similar modeling issues are also addressed in the book.

1.2.6 Computational Methods

Following the theoretical work on establishing the optimal control policies under various conditions, numerical procedures for computing optimal policies have also been developed. Although the structure of (s, S) policies is fairly simple, the computation of an (s, S) policy turns out to be difficult. Traditionally, two main approaches were explored in computing the actual (s, S) values. The first is based on the dynamic programming formulation of the problem. The second approach, known as stationary analysis, involves the determination of the limiting distribution of the inventory position under an (s, S) policy. Using this distribution, an expected cost function is constructed with s and S as decision variables; (see, e.g., Veinott and Wagner (1965)). Minimization of this function determines the optimal (s, S) policy. A computationally more efficient approach is the one developed by Zheng and Federgruen (1991).

The literature reviewed here is intended to provide the readers with a basic idea about which models are relevant to this book. How models presented in different chapters of this book relate to the literature will be discussed in those chapters.

1.3. Examples of Markovian Demand Models

The classical and best known stochastic inventory models assume the demand in different time periods to be independent and often even identically distributed. Nonetheless, demand forecasters have known all along that the demand over time for most products does not form a sequence of independent random variables. Most forecasting techniques either rely on trends and autocorrelation in the sequence of demands or on the fact that future demands are correlated to some observable early indicator. A specific structure of a Markov-modulated demand process is used in this book in an attempt to formalize these dependencies and allow for them when making inventory decisions. Below are a few examples of such indicators and demand processes consistent with our models.

The influence of weather on the demands for many products is one of the first things that come to mind. The demand for home-heating oil in winter depends heavily on the temperature outside. Likewise, the demand for electricity in summer increases as the temperatures rise and vice versa. Different temperature ranges can be thought of as different states of the world determining the probability distribution of demand. Once the temperature is known, the demand is still uncertain but two things have changed; first, knowing the ambient temperature, demand can likely be predicted more accurately, and therefore the demand distribution will have lower variance. Second, since temperatures in adjacent time periods are correlated, those correlations can be exploited to predict future temperatures and demands. The transition matrices of the temperatures in a geographic area can be estimated from the historical weather data. Also note that these matrices could be nonstationary over time, i.e., they may be different in different months. For more weather related examples, the demand for insecticides in an area may be positively correlated to the amount of total rainfall in the preceding week. A good indicator for gasoline demand at gas stations along stretches of highway I-80 would be the weather forecast for the Lake Tahoe region. Here, the demand could be a high-level categorization of the forecast; for example, pleasant, mediocre, or unpleasant.

When it comes to estimating the demand for consumable materials like ink or toner cartridges for a specific type of printer, it is obvious that the number of such printers currently in use, often referred to as the installed base, is an important piece of information. Therefore, the sales volume combined with data about typical usage patterns of the products requiring the same consumables can be used to provide the state information of the demand. The transition matrix of demand states can also be estimated based on the projected life spans of the relevant products. Similarly, in the case of demand for spare parts, the number of installed units (for example, servers) in a given age range should be a good choice to use as the demand state, since the failure rate of each of the items is usually related to its age.

Demand for replacement units under warranty also exhibits a similar pattern that is directly associated with the number of new units (for example, a specific model of a digital camera or a mainframe computer) sold recently. Many electronic devices show high "infant mortality." A large number of those that fail are either defective on arrival or fail shortly after being put to use. For these reasons, the number of replacement units requested under warranty is positively correlated to the recent sales volume of the product, a reasonable choice for the demand state.

The product life cycle is another important indicator for product demand. Typically, sales volume starts low in the introductory stage, increases significantly in the growth stage, reaches the peak in the mature stage, and then falls in the decline stage. It is a natural and realistic way to model the transitions of a product life cycle from one stage to the next as a Markov chain.

In the setting of competitive market environment, the introduction of a competing product into the market can be modeled as a Markov process with the state variable representing the number of competing products. A product may initially be the market leader when there are no competitive products at the time of its introduction. However, its market share is likely to diminish as competitive products are introduced over time.

In the IT industry, the introduction of new software is often a catalyst for hardware demand. For example, demand for memory upgrades on notebook computers will be higher after the introduction of a new version of software such as a new version of Microsoft Windows operating system. In this case, the introduction of the new Windows operating system can be used as the demand state when a Markovian model is adopted for modeling the demand of an affected hardware product.

Seasonality is a common factor that affects product demand. For example, the demand for many consumer electronics products like notebook computers or inkjet printers exhibits strong seasonal patterns. Demand for these products is the highest during "back to school" season (August, September) and the holiday season (November, December). During the remainder of the year, the monthly demand is lower and more stable. Seasonal demand is also typical in commercial sales. It is well known that IT related purchases and capital investments (servers, large laser printers, storage devices, etc.) follow a so-called "hockey stick" curve. Demand tends to be highly skewed towards the end of a month or a calendar quarter. This is thought to be the result of the spend-it-or-loose-it regime of corporate budgeting, as well as the structure of sales incentives. This seasonality of demand can be captured by a degenerate form of a Markov chain with the transition from one state to the next being deterministic (or almost deterministic). See Section 2.7 of Chapter 2 for further discussion on this type of model.

The examples given above illustrate that Markovian demand models can capture the relationship between the demand and the associated environmental factors. Certainly, these models are more realistic and more flexible representations of the demand process than i.i.d. demand models or nonstationary independent demand models. On the other hand, the inventory models allowing for Markovian demands are more complex than the classical inventory models especially in terms of data and computational requirements. However, inventory decisions have tremendous business consequences in terms of cost as well as customer satisfaction. Because of this, the additional effort required to use these models will easily be justified in many cases by the value of improved decisions resulting from them.

1.4. Contributions

One of the most important developments in the inventory theory has been to show that (s, S) policies (with base-stock policy as a special case) are optimal for a class of dynamic inventory models with random periodic demands and fixed ordering costs.

Dynamic inventory models can be formulated in a continuous-time setting or a discrete-time setting. The models presented in this book are of the latter type.

The main contribution of the book is to generalize a class of inventory models (with fixed ordering costs) to allow for demands that depend on an exogenous Markov process or a controlled Markov process, and to show that the optimal policies are of (s, S)-type. In our models, the distribution of demands in successive periods is dependent on a Markov chain. Special cases of such Markovian demand models include the case of cyclic or seasonal demand. Finite horizon, as well as stationary and nonstationary infinite horizon, problems are considered. Some constraints commonly encountered in practice, namely no-ordering periods, finite storage capacities, and service levels, are also treated. We show that (s, S)-type policies are also optimal for the generalized models as well as their extensions.

We provide rigorous analyses under various modeling assumptions: discounted cost models with backlog or lost sales assumptions, average cost models (also with backlog or lost sales assumptions), models with demand controlled by a Markov decision process (MDP), and models with cost functions of polynomial growth. Required results on the existence of optimal solutions, as well as verification theorems, are established for the cases studied.

1.5. Plan of the Book

This book consists of six parts. Part I (consisting of just Chapter 1) is an introduction to the book. Part II (Chapters 2-4) covers the models with the discounted cost criterion. The models with the average cost criterion are presented in Part III (Chapters 5-7). Part IV (Chapters 8 and 9) collects miscellaneous results that are not covered in the previous chapters. The concluding remarks are discussed in Part V. Appendices are included in Part VI.

Part II starts with Chapter 2, where a discounted cost model with the full backlog assumption is introduced. The model is a generalization of classical inventory models with fixed costs that exhibit (s, S) policies. We model the demand process in a way that the demand distributions in successive periods depend upon the state of an underlying Markov chain. A dynamic programming formulation is used to provide the results on the uniqueness of the solution and the existence of an optimal feedback policy. Also described in the chapter are some new properties of K-convex functions, which provide the technical results needed for the analysis in the rest of the chapter. The optimality of an (s, S)-type policy is first shown in a finite horizon model, with s and S dependent on the demand state of the current time period and on the time remaining. This result is extended to a nonstationary infinite horizon version of the model. Extensions with more realistic modeling assumptions are also presented in this chapter.

Chapter 3 treats the discounted infinite horizon inventory model involving fixed cost in Chapter 2 to allow for unbounded demand and costs with polynomial growth. Finite horizon problems, as well as stationary and nonstationary discounted cost infinite horizon problems, are addressed. Existence of optimal Markov or feedback policies is established with unbounded Markovian demand, ordering costs that are l.s.c., and inventory/backlog (or surplus) costs that are l.s.c. with polynomial growth. Furthermore, optimality of state-dependent (s, S) policies is proved when the ordering cost consists of fixed and proportional cost components and the surplus cost is convex.

Chapter 4 extends the results developed in Chapter 2 to the lost sales case, where demand that is not satisfied on time is lost. The lost sales case is typically more difficult to analyze than its backlog counterpart. Generally it requires a different treatment and sometimes additional assumptions to establish the same type of results as in the backlog case. A new K-convexity result associated with the cost functions in the lost sales case is presented in this chapter. Based on this new result, we are able to show that a state-dependent (s, S) policy is optimal for the Markovian demand inventory models with the lost sales assumption in both finite horizon and infinite horizon formulations. Extensions that incorporate various realistic features and constraints are also developed in the chapter.

Part III, consisting of Chapters 5, 6 and 7, analyzes the average cost models. Chapter 5 deals with a long-run average cost version of the Markovian demand model treated in Chapter 2 with the backlog assumption, fixed ordering cost, and convex inventory cost. We develop a vanishing discount approach to establish the average cost optimality equation and provide the associated verification theorem to show that a state-dependent (s, S) policy is optimal for the average cost model.

Chapter 6 is devoted to studying the long-run average cost version of the model presented in Chapter 3 with unbounded Markovian demands, ordering cost that are l.s.c., and inventory costs that are l.s.c. and of polynomial growth. Finite horizon problems and stationary long-run average cost problems are addressed.

Chapter 7 provides an analysis of the lost sales version of the problem described in Chapter 5. Similar to the approach used in Chapter 5, we examine the asymptotic behavior of the differential discounted value function as the discount rate approaches zero. Then, we establish the average cost optimality equation using the vanishing discount approach. Also included is the proof of the verification theorem as well as the optimality proof for a state-dependent (s, S) policy in this setting.

In Part IV, two additional topics that are not covered by the previous chapters are discussed. In Chapter 8, we present a MDP based model for a joint inventory/promotion decision problem. The state variable of the MDP represents the demand state brought about by changing environmental factors as well as promotion decisions. We show that the optimal joint inventory/promotion decision can be characterized by policy of a simple form under certain conditions, where the promotion decision follows a threshold policy and the inventory decision follows a base-stock policy, with both policies dependent on the demand state.

Chapter 9 revisits the classical papers of Iglehart (1963b) and Veinott and Wagner (1965) devoted to stochastic inventory problems with the criterion of long-run average cost minimization. We indicate some of the assumptions that are implicitly used without verification in their stationary distribution approach to the problems, and provide the missing (nontrivial) verification. In addition to completing their analysis, we examine the relationship between the stationary distribution approach and the dynamic programming approach to the average cost stochastic inventory problems.

We conclude the book with Part V consisting of Chapter 10, where conclusions and open research problems are described.

Part VI consists of Appendices. These Appendices provide some background material as well as the technical results that are used in the book.

The relationships between different chapters are shown in Figure 1.1.

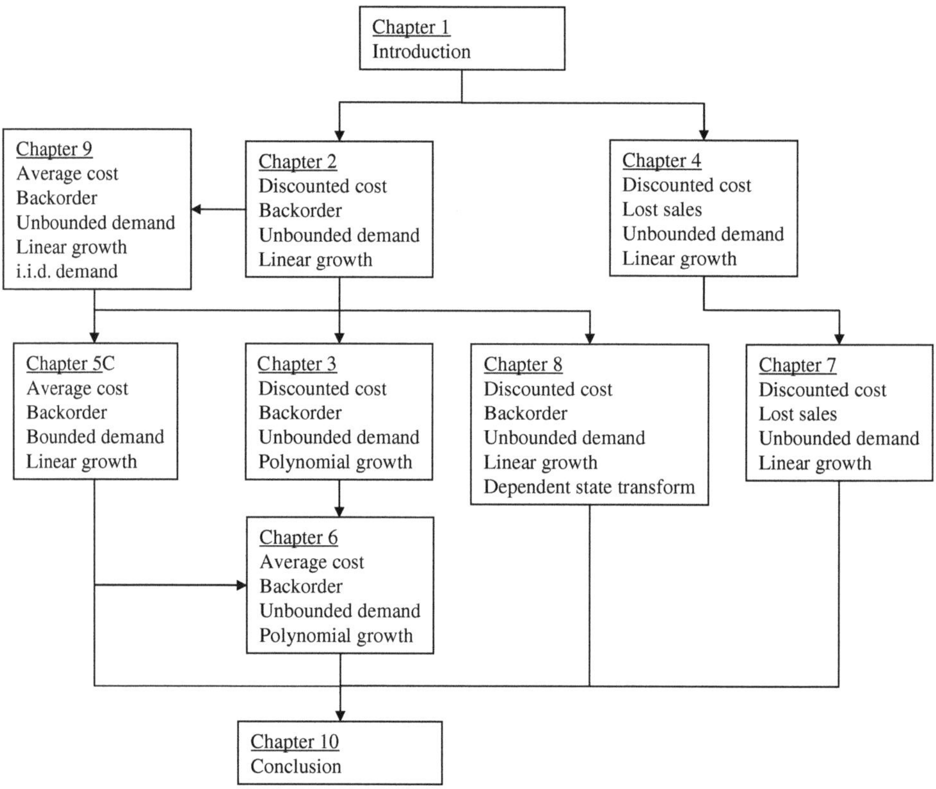

Figure 1.1. Relationships between different chapters.

An extensive bibliography, author index, subject index, and the copyright permissions follow the Appendices.

II

DISCOUNTED COST MODELS

Chapter 2

DISCOUNTED COST MODELS WITH BACKORDERS

2.1. Introduction

One of the most important developments in the inventory theory has been to show that (s, S) policies are optimal for a class of dynamic inventory models with random periodic demands and fixed ordering costs. Under an (s, S) policy, if the inventory level at the beginning of a period is less than the reorder point s, then a sufficient quantity must be ordered to achieve an inventory level S, the order-up-to level, upon replenishment. There are a number of papers in the literature devoted to proving the optimality of (s, S) policies under a variety of assumptions. However, in real-life inventory problems, some of these assumptions do not hold. It is our purpose to relax these assumptions toward realism and still demonstrate the optimality of (s, S)-type policies.

The nature of the demand process is an important assumption in stochastic inventory models. With possible exceptions of Karlin and Fabens (1960) and Iglehart and Karlin (1962), classical inventory models have assumed demand in each period to be a random variable independent of demands in other periods and of environmental factors other than time. However, as elaborated in Song and Zipkin (1993), many randomly changing environmental factors, such as fluctuating economic conditions and uncertain market conditions in different stages of a product life cycle, can have a major effect on demand. For such situations, the Markov chain approach provides a natural and flexible alternative for modeling the demand process. In such an approach, environmental factors are represented by the *demand state* or the *state-of-the-world* of a Markov process, and demand in a period is a random variable with its distribution function dependent on the demand state in that period.

D. Beyer et al., *Markovian Demand Inventory Models*, International Series in Operations Research & Management Science 108,
DOI 10.1007/978-0-387-71604-6_2, © Springer Science+Business Media, LLC 2010

Furthermore, the demand state can also affect other parameters of the inventory system such as the cost functions.

Another feature that is not usually treated in the classical inventory models, but is often observed in real life is the presence of various constraints on ordering decisions and inventory levels. For example, there may be periods, such as weekends and holidays, during which deliveries cannot take place. Also, the maximum inventory that can be accommodated is often limited by a finite storage space. On the other hand, one may wish to keep the amount of inventory above a certain level to reduce the chance of a stockout, and ensure satisfactory service to customers.

While some of these features are dealt with in the literature in a piecemeal fashion, we will formulate a sufficiently general model that has models with one or more of these features as special cases and still retain the optimal policy to be of (s, S)-type. Thus, our model considers more general demands, costs, and constraints than most of the fixed cost inventory models in the literature.

The plan of this chapter is as follows. The next section contains a review of relevant models and how our model relates to them. In Section 2.3, we develop a general finite horizon inventory model with a Markovian demand process. In Section 2.4, we state the dynamic programming equations for the problem and the results on the uniqueness of the solution and the existence of an optimal feedback or Markov policy. In Section 2.5, we use some properties of K-convex functions, derived in Appendix C, to show that the optimal policy for the finite horizon model under consideration is still of (s, S)-type, with the policy parameters s and S dependent on the demand state and the time remaining. The nonstationary infinite horizon version of the model is examined in Section 2.6. The cyclic demand case is treated in Section 2.7. The analysis of models incorporating no-ordering periods and those with the shelf capacity and service level constraints is presented in Section 2.8. Section 2.9 concludes the chapter.

2.2. Review of the Related Literature

Classical papers on the optimality of (s, S) policies in dynamic inventory models with stochastic demands and fixed setup costs include those of Arrow *et al.*(1951), Dvoretzky *et al.* (1953), Karlin (1958a), Scarf (1960), Iglehart (1963b), and Veinott (1966). Scarf develops the concept of K-convexity and uses it to show that (s, S) policies are optimal for finite horizon inventory problems with fixed ordering costs. That a stationary (s, S) policy is optimal for the stationary infinite horizon problem is proved by Iglehart (1963b). Furthermore, Bensoussan *et al.* (1983) provide a rigorous formulation of the problem with nonstationary but stochastically independent demand. They also deal with the issue of the existence of optimal feedback policies, along with a proof of the optimality of an (s, S)-type policy in the nonstationary finite, as well as infinite horizon cases.

The effect of a randomly changing environment in inventory models with fixed costs received only limited attention in the early literature. Karlin and Fabens (1960) have introduced a Markovian demand model similar to ours. They indicate that given the Markovian demand structure in their model, it appears reasonable to postulate an inventory policy of (s, S)-type with a different set of critical numbers for each demand state. But they consider the analysis to be complex, and concentrate instead on optimizing only over the restricted class of ordering policies, each characterized by a single pair of critical numbers, s and S, irrespective of the demand state.

Song and Zipkin (1993) present a continuous-time, discrete-state formulation with a Markov-modulated Poisson demand and with linear costs of inventory and backlogging. They show the optimality of a state-dependent (s, S) policy when the ordering cost consists of both a fixed cost and a linear variable cost. An algorithm for computing the optimal policy is also developed using a modified value iteration approach.

The basic model presented in the next section extends the classical Karlin and Fabens model in two significant ways. It generalizes the cost functions that are involved and it optimizes over the natural class of all history-dependent ordering policies. In relation to Song and Zipkin (1993), we consider more general demands (see Remark 2.3) and state-dependent convex inventory/backlog costs without a standard assumption made in the literature on backlog and purchase costs; (see Remarks 2.2 and 2.1). The nonstationary infinite horizon model extends Bensoussan *et al.* (1983) to allow for Markovian demands and more general asymptotic behavior on the shortage cost as the shortage becomes large; (see Remark 2.1).

2.3. Formulation of the Model

Let us consider an inventory problem over a finite number of periods $\langle n, N\rangle = \{n, n+1, \ldots, N\}$ and an initial inventory of x units at the beginning of period n, where n and N are any given integers satisfying $0 \le n \le N < \infty$. The demand in each period is assumed to be a random variable defined on a given probability space $(\Omega, \mathcal{F}, \mathsf{P})$ and not necessarily identically distributed. More specifically, the demand distributions in successive periods are defined as below.

Consider a finite collection of demand states $\mathbb{I} = \{1, 2, \ldots, L\}$, and let i_k denote the demand state in period k. We assume that i_k, $k \in [n, N]$, with known initial demand state i_n is a Markov chain over $\mathbb{I}$ with the transition matrix $P = \{p_{ij}\}$. Thus,

$$0 \le p_{ij} \le 1,\ i \in \mathbb{I}, j \in \mathbb{I}, \text{ and } \sum_{j=1}^{L} p_{ij} = 1,\ i \in \mathbb{I}.$$

Let a nonnegative random variable ξ_k denote the demand in a given period k, $k = 0, \ldots, N-1$. Demand ξ_k depends only on period k and the demand state in that period, by which we mean that it is independent of past demand states and past demands. We denote its cumulative probability distribution by $\Phi_{i,k}(x)$, when the demand state $i_k = i$. In the following period, if the state changes to state j, which happens with probability p_{ij}, then the demand distribution is $\Phi_{j,k+1}$ in that period. We further assume that for a positive constant D,

$$\mathsf{E}(\xi_k | i_k = i) = \int_0^\infty x d\Phi_{i,k}(x) \le D < \infty,\ k = 0, \ldots, N-1,\ i \in \mathbb{I}. \tag{2.1}$$

This is not a very restrictive assumption from an applied perspective.

We denote by

$$\begin{aligned} &\mathcal{F}_l^k, \text{ the } \sigma\text{-algebra generated by } \{i_l, \ldots, i_{k-1}, i_k; \xi_l, \ldots, \xi_{k-1}\}, \\ &\hphantom{\mathcal{F}_l^k, \text{ the } \sigma\text{-algebra generated by }} 0 \le l \le k \le N, \\ &\mathcal{F}^k = \mathcal{F}_0^k. \end{aligned} \tag{2.2}$$

Since $i_k,\ k = 1, \ldots, N$, is a Markov chain and ξ_k depends only on i_k, we have

$$\begin{aligned} \mathsf{E}(\xi_k | \mathcal{F}^k) &= \mathsf{E}(\xi_k | i_0, i_1, \ldots, i_k; \xi_0, \xi_1, \ldots, \xi_{k-1}) \\ &= \mathsf{E}(\xi_k | i_k). \end{aligned} \tag{2.3}$$

An admissible decision (ordering quantities) for the problem on the interval $[n, N]$ with initial state $i_n = i$ can be denoted as

$$U = (u_n, \ldots, u_{N-1}), \tag{2.4}$$

where u_k is a nonnegative $\mathcal{F}_n^k$-measurable random variable. In simpler terms, this means that decision u_k depends only on the past information. Note that since i_n is known in period n, $\mathcal{F}_n^n = (\Omega, \emptyset)$, and hence

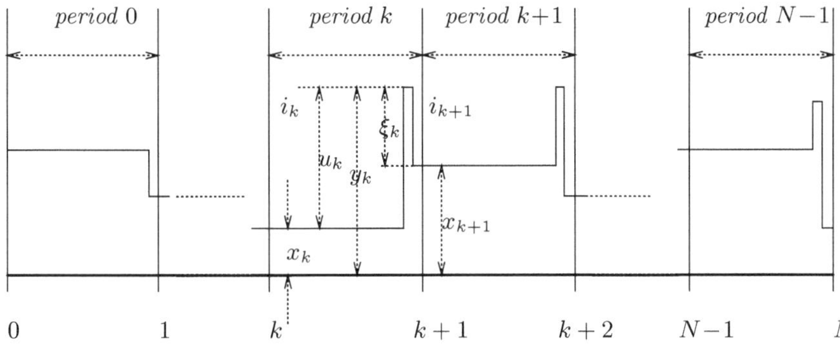

Figure 2.1. Temporal conventions used for the discrete-time inventory problem.

u_n is deterministic. Moreover, it should be emphasized that this class of admissible decisions is larger than the class of admissible feedback policies. Let us also denote a policy of not ordering anything at all as $\mathbf{0} = (0, \ldots, 0)$.

Ordering quantities are decided upon at the beginning of each period. Demand in each period is supposed to occur at the end of the period after the order has been delivered; (see Figure 2.1 for the temporal conventions used). Unsatisfied demand is carried forward as backlog. The inventory balance equations are defined by

$$\begin{cases} x_{k+1} = x_k + u_k - \xi_k, \ k = n, \ldots, N-1, \\ x_n = x, \ \text{ initial inventory level,} \\ i_k, k = n, \ldots, N, \ \text{Markov chain with transition matrix } P, \\ i_n = i, \ \text{ initial state,} \end{cases}$$

where x_k is the surplus level at the beginning of period k, u_k is the quantity ordered at the beginning of period k, i_k is the demand state in period k, and ξ_k is the demand in period k. Note that $x_k > 0$ represents an inventory of x_k and $x_k < 0$ represents a backlog (or shortage) of $-x_k$. Also, the initial state i and the initial inventory level x are assumed to be arbitrarily given.

Furthermore, we specify the relevant costs and the assumptions they satisfy.

(i) The purchase or production cost is expressed as

$$c_k(i, u) = K_k^i \mathbb{I}_{u>0} + c_k^i u, \quad k \in \langle 0, N-1 \rangle, \tag{2.5}$$

where the fixed ordering costs are $K_k^i \geq 0$, the variable costs are $c_k^i \geq 0$, and $\mathbb{I}_{u>0}$ equals 1 when $u > 0$ and equals 0 when $u \leq 0$.

(ii) The surplus (or inventory/backlog) cost functions $f_k(i,\cdot)$ are convex, and they are asymptotically linear, i.e.,

$$f_k(i,x) \leq C(1+|x|) \quad \text{for some } C>0, \quad k \in \langle 0, N\rangle. \tag{2.6}$$

The objective function to be minimized is the expected value of all the costs incurred during the interval $\langle n, N\rangle$ with $i_n = i$ and $x_n = x$, i.e.,

$$J_n(i,x;U) = \mathsf{E}\left\{\sum_{k=n}^{N-1} [c_k(i_k,u_k) + f_k(i_k,x_k)] + f_N(i_N,x_N)\right\}, \tag{2.7}$$

where $U = (u_n,\ldots,u_{N-1})$ is a *history-dependent* or *nonanticipative* admissible decision (order quantities) for the problem and $u_N = 0$. The inventory balance equations are given by

$$x_{k+1} = x_k + u_k - \xi_k,\ k \in \langle n, N-1\rangle. \tag{2.8}$$

Finally, we define the value function for the problem over $\langle n, N\rangle$ with $i_n = i$ and $x_n = x$ to be

$$v_n(i,x) = \inf_{U\in\mathcal{U}} J_n(i,x;U), \tag{2.9}$$

where $\mathcal{U}$ denotes the class of all admissible decisions. Note that the existence of an optimal policy is not required to define the value function. Of course, once the existence is established, the “inf” in (2.9) can be replaced by “min”.

2.4. Dynamic Programming and Optimal Feedback Policy

In this section we develop the dynamic programming equations satisfied by the value function. We then provide a verification theorem (Theorem 2.2), which states that the cost associated with the feedback or Markov policy obtained from the solution of the dynamic programming equations, equals the value function of the problem on $\langle 0, N\rangle$.

Let $\mathbb{B}_0$ denote the class of all continuous functions from $\mathbb{I}\times\mathbb{R}$ into $\mathbb{R}^+$ and the pointwise limits of sequences of these functions; (see Feller (1971)). Note that it includes piecewise-continuous functions. Let $\mathbb{B}_1$ be the space of functions in $\mathbb{B}_0$ that are of linear growth, i.e., for any $b \in \mathbb{B}_1$, $0 \leq b(i,x) \leq C_b(1+|x|)$ for some $C_b > 0$. Let $\mathbb{C}_1$ be the subspace of functions in $\mathbb{B}_1$ that are uniformly continuous with respect to $x \in \mathbb{R}$. For any $b \in \mathbb{B}_1$, we define the norm

$$\| b \| = \max_i \sup_x \frac{b(i,x)}{1+|x|}$$

and the operator

$$\begin{aligned} F_{k+1}b(i,y) &= \mathsf{E}[b(i_{k+1}, y-\xi_k)|i_k = i] \\ &= \sum_{j=1}^{L} \{\mathsf{P}(i_{k+1} = j | i_k = i)\mathsf{E}[b(j, y-\xi_k)|i_k = i]\} \\ &= \sum_{j=1}^{L} p_{ij} \int_0^\infty b(j, y-\xi) d\Phi_{i,k}(\xi). \end{aligned} \tag{2.10}$$

In addition to Assumptions (i) and (ii) on costs, we also require that for $k = 0, 1, \ldots, N-1$,

$$c_k^i x + F_{k+1}(f_{k+1})(i,x) \to +\infty \text{ as } x \to \infty. \tag{2.11}$$

REMARK 2.1 Condition (2.11) means that either the unit ordering cost $c_k^i > 0$ or the expected holding cost $F_{k+1}(f_{k+1})(i,x) \to +\infty$ as $x \to \infty$, or both. Condition (2.11) is borne out of practical considerations and is not very restrictive. In addition, it rules out such unrealistic trivial cases as the one with $c_k^i = 0$ and $f_k(i,x) = 0, x \geq 0$, for each i and k, which implies ordering an infinite amount whenever an order is placed. The condition generalizes the usual assumptions made by Scarf (1960) and others that the unit inventory carrying cost $h > 0$. Furthermore, because of an essential asymmetry between the inventory side and the backlog side we need not impose a condition like (2.11) on the backlog side assumed in Bensoussan *et al.* (1983) and Bertsekas (1976). Whereas we can order any number of units to decrease backlog or build inventory, it is not possible to sell anything more than the demand in order to decrease inventory or increase backlog. If it were possible, then the condition like (2.11) as $x \to -\infty$ would be needed to make backlog more expensive than the revenue obtained by sale of units, asymptotically. In the special case of stationary linear backlog costs, this would imply $p > c$ (or $p > \alpha c$ if costs are discounted at the rate $\alpha, 0 < \alpha \leq 1$), where p is the unit backlog cost. But since sales in excess of demand are not allowed, we are able to dispense with the condition like (2.11) on the backlog side or the standard assumption $p > c$ (or $p > \alpha c$) as in Scarf (1960) and others, or the strong assumption $p > \alpha c^i$ for each i as in Song and Zipkin (1993).

Using the principle of optimality, we can write the following dynamic programming equations for the value function

$$\begin{cases} v_n(i,x) &= f_n(i,x) + \inf_{u\geq 0} \{c_n(i,u) \\ & \quad + \mathsf{E}[v_{n+1}(i_{n+1}, x+u-\xi_n)|i_n = i]\} \\ &= f_n(i,x) + \inf_{u\geq 0} \{c_n(i,u) + F_{n+1}(v_{n+1})(i,x+u)\}, \\ & \qquad\qquad n \in \langle 0, N-1 \rangle, \\ v_N(i,x) &= f_N(i,x). \end{cases} \tag{2.12}$$

The next two theorems are fundamental. Together, they prove that (i) the dynamic programming equations have a solution in an appropriate space, (ii) the infima in these equations are attained, (iii) these infima provide an optimal feedback control within the class of admissible controls, and (iv) the solution of the equations is the value function. Theorem 2.1 proves (i) and (ii). Theorem 2.2 takes the solution of the dynamic programming equations and the infima obtained in Theorem 2.1 and goes on to prove (iii) and (iv). Its proof uses the property (vii) of conditional expectations given in Appendix B.2. Note, furthermore, that from (iv) it follows that the solution of the dynamic programming equations is unique in the defined space.

THEOREM 2.1 *The dynamic programming equations* (2.12) *define a sequence of functions in* $\mathbb{C}_1$. *Moreover, there exists a function* $\hat{u}_n(i,x)$ *in* $\mathbb{B}_0$, *which provides the infimum in* (2.12) *for any* x.

Proof. We proceed by induction. By Assumption (ii) on the function f_N and Theorems C.2.1 and A.1.3, v_N is in $\mathbb{C}_1$. Now, assume that $v_{n+1}(i,x)$ belongs to $\mathbb{C}_1$ for $n < N$. Consider points x such that $|x| \leq M$. It follows from (2.12) that $v_n(i,x) \geq f_n(i,x)$ for all $n, i,$ and x. Let

$$B_{n,i}^M = \max_{|x|\leq M} \{c_n(i,0) + F_{n+1}(v_{n+1})(i,x)\}.$$

We know that $B_{n,i}^M < \infty$ because $v_{n+1} \in \mathbb{C}_1$. Let $y = x + u$. Then, we have

$$\begin{aligned} c_n(i,u) + F_{n+1}(v_{n+1})(i,x+u) &\geq K_n^i + c_n^i(y-x) + F_{n+1}(f_{n+1})(i,y) \\ &\geq K_n^i - c_n^i M + c_n^i y + F_{n+1}(f_{n+1})(i,y). \end{aligned}$$

Because of (2.11), there is a constant $\bar{u}_{n,i}^M$ such that for all $y > \bar{u}_{n,i}^M - M$, we have

$$c_n^i y + F_{n+1}(f_{n+1})(i,y) > B_{n,i}^M + c_n^i M - K_n^i.$$

Since $y > \bar{u}^M_{n,i} - M$ is implied by $u > \bar{u}^M_{n,i}$, we have for all $u > \bar{u}^M_{n,i}$,

$$c_n(i,u) + F_{n+1}(v_{n+1})(i, x+u) > B^M_{n,i} \geq c_n(i,0) + F_{n+1}(v_{n+1})(i,x).$$

Consequently, any $u > \bar{u}^M_{n,i}$ cannot be the infimum in (2.12). Therefore in (2.12), we can restrict u by the constraint $0 \leq u \leq \bar{u}^M_{n,i}$ for all the points x satisfying $|x| \leq M$, without loss of optimality.

Since the function

$$\psi_n(i,x;u) = c_n(i,u) + F_{n+1}(v_{n+1})(i,x+u)$$

is l.s.c. and bounded from below, its minimum over a compact set is attained. Moreover, from the Selection Theorem A.1.7, we know that there exists a Borel function $\hat{u}^M_n(i,x)$ such that

$$\psi_n(i,x;\hat{u}^M_n(i,x)) = \inf_{0 \leq u \leq \bar{u}^M_{n,i}} \psi_n(i,x;u), \ \forall x.$$

With the definition

$$\hat{u}_n(i,x) = \hat{u}^M_n(i,x) \ \text{ for } M-1 < |x| \leq M,$$

we obtain a Borel function such that

$$\psi_n(i,x,\hat{u}_n(i,x)) = \inf_{u \geq 0} \psi_n(i,x,u), \ \forall x.$$

Now for $|x_1 - x_2| \leq \delta$, we have

$$\begin{aligned}
&|\psi_n(i,x_1,u) - \psi_n(i,x_2,u)| \\
&\quad = |F_{n+1}(v_{n+1})(i,x_1+u) - F_{n+1}(v_{n+1})(i,x_2+u)| \\
&\quad \leq \sum_{j=1}^{L} p_{ij} \sup_{|x_1-x_2|\leq\delta} |v_{n+1}(j,x_1) - v_{n+1}(j,x_2)|,
\end{aligned}$$

from which together with (2.5), (2.6) and (2.12), it follows easily that $v_n(i,x)$ is uniformly continuous in x. Since

$$\inf_{u\geq 0} \psi_n(i,x,u) \leq c_n(i,0) + \| F_{n+1} \| \| v_{n+1} \| (1+|x|),$$

we can use (2.12) and (2.6) to conclude that $v_n(i,x) \in \mathbb{C}_1$. □

To solve the problem of minimizing $J_0(i,x;U)$, we use $\hat{u}_n(i,x)$ of Theorem 2.1 to define

$$\begin{cases} \hat{u}_n = \hat{u}_n(i_n, \hat{x}_n), \ n \in \langle 0, N-1 \rangle \ \text{ with } i_0 = i, \\ \hat{x}_{n+1} = \hat{x}_n + \hat{u}_n - \xi_n, \ n \in \langle 0, N-1 \rangle \ \text{ with } \hat{x}_0 = x. \end{cases} \tag{2.13}$$

THEOREM 2.2 (Verification Theorem) *The decision* $\hat{U}=(\hat{u}_0,\hat{u}_1,\ldots,\hat{u}_{N-1})$ *is optimal for the problem* $J_0(i,x;U)$. *Moreover,*

$$v_0(i,x) = \min_{U\in\mathcal{U}} J_0(i,x;U). \tag{2.14}$$

Proof. Let $U = (u_0,\ldots,u_{N-1})$ be any admissible decision. Without loss of generality, we may assume that $\mathsf{E}c_n(i_n,u_n) < \infty$, $\mathsf{E}f_n(i_n,x_n) < \infty$, $n \in \langle 0, N-1\rangle$ and $\mathsf{E}f_N(i_N,x_N) < \infty$. Otherwise, $J_0(i,x;U) = \infty$ and U cannot be optimal since $J_0(i,x;\mathbf{0}) < \infty$ in view of (2.1) and Assumptions (i) and (ii).

Because $v_N(i_N,x_N) = f_N(i_N,x_N)$, it follows that $\mathsf{E}v_N(i_N,x_N) < \infty$. We proceed by induction. Assume that $\mathsf{E}v_{n+1}(i_{n+1},y_{n+1}) < \infty$. Next, using (2.8), the property (B.2-vii) of conditional expectations, the Markovian property (2.3), the independence assumption of ξ_n, and the notation (2.10), we obtain

$$\begin{aligned}
&\mathsf{E}\{v_{n+1}(i_{n+1},x_{n+1})|i_0,\ldots,i_n,\xi_0,\ldots,\xi_{n-1}\}\\
&\quad= \mathsf{E}\{v_{n+1}(i_{n+1},x_n+u_n-\xi_n)|i_0,\ldots,i_n,\xi_0,\ldots,\xi_{n-1}\}\\
&\quad= \mathsf{E}\{v_{n+1}(i_{n+1},y-\xi_n)|i_0,\ldots,i_n,\xi_0,\ldots,\xi_{n-1}\}_{y=x_n+u_n}\\
&\quad= \mathsf{E}\{v_{n+1}(i_{n+1},y-\xi_n)|i_n\}_{y=x_n+u_n}\\
&\quad= F_{n+1}(v_{n+1})(i_{n+1},y)_{y=x_n+u_n}\\
&\quad= F_{n+1}(v_{n+1})(i_{n+1},x_n+u_n) \text{ a.s.}
\end{aligned} \tag{2.15}$$

Now using (2.12), since U is admissible but not necessarily optimal, we can assert that

$$v_n(i_n,x_n) \le f_n(i_n,x_n) + c_n(i_n,u_n) + F_{n+1}(v_{n+1})(i_{n+1},x_n+u_n) \text{ a.s.,}$$

and from the relation (2.15), we can derive

$$\begin{aligned}
v_n(i_n,x_n) \;\le\; & f_n(i_n,x_n) + c_n(i_n,u_n)\\
&+\mathsf{E}\{v_{n+1}(i_{n+1},x_{n+1})|i_0,\ldots,i_n,\xi_0,\ldots,\xi_{n-1}\} \text{ a.s.}
\end{aligned}$$

By taking the expectation of both sides of the above inequality, we obtain

$$\mathsf{E}v_n(i_n,x_n) \le \mathsf{E}(f_n(i_n,x_n) + c_n(i_n,u_n)) + \mathsf{E}(v_{n+1}(i_{n+1},x_{n+1})). \tag{2.16}$$

It follow from (2.16) that $\mathsf{E}v_n(i_n,x_n) < \infty$ and, therefore, (2.16) holds for all $n \in \langle 0, N\rangle$. Summing from 0 to $N-1$, we get

$$v_0(i,x) \le J_0(i,x;U). \tag{2.17}$$

Now consider the decision $\hat{U}$. From the definition of $\hat{u}_n(i_n, x)$ as the Borel function that attains the infimum in (2.12), and proceeding as above, we can also obtain

$$\mathsf{E}v_n(i_n, \hat{y}_n) = \mathsf{E}(f_n(i_n, \hat{y}_n) + c_n(i_n, \hat{u}_n)) + \mathsf{E}(v_{n+1}(i_{n+1}, y_{n+1})).$$

Note that $\hat{x}_0 = x$ is deterministic and $v_0(i, x) \in \mathbb{C}_1$. Thus, $\mathsf{E}v_0(i_0, \hat{x}_0) = v_0(i, x) < \infty$, and we can prove recursively that $\mathsf{E}c_n(i_n, \hat{u}_n) < \infty$, $n \in \langle 0, N-1 \rangle$ and $\mathsf{E}f_n(i_n, \hat{x}_n) < \infty$, $\mathsf{E}v_n(i_n, \hat{x}_n) < \infty$, $n \in \langle 0, N \rangle$. Adding up for n from 0 to $N-1$, it follows that

$$v_0(i, x) = J_0(i, x; \hat{U}).$$

This and the inequality (2.17) complete the proof. □

Taken together, Theorems 2.1 and 2.2 establish the existence of an optimal feedback policy. This means that there exists a policy in the class of admissible policies whose objective function value equals the value function defined by (2.9), as well as a Markov (or feedback) policy which gives the same objective function value. Furthermore, the solution $v_0(i, x)$ obtained in Theorem 2.1 is the value function.

2.5. Optimality of (s, S)-type Ordering Policies

We impose an additional condition on the costs under which the optimal feedback policy $\hat{u}_n(i, x)$ turns out to be an (s, S)-type policy. For $n \in \langle 0, N-1 \rangle$ and $i \in \mathbb{I}$, let

$$K_n^i \geq \bar{K}_{n+1}^i \equiv \sum_{j=1}^{L} p_{ij} K_{n+1}^j \geq 0. \tag{2.18}$$

Remark 2.2 Condition (2.18) means that the fixed cost of ordering in a given period with demand state i should be no less than the expected fixed cost of ordering in the next period. The condition is a generalization of the similar conditions used in the standard models. It includes the cases of the constant ordering costs ($K_n^i = K$, $\forall i, n$) and the nonincreasing ordering costs ($K_n^i \geq K_{n+1}^j$, $\forall i, j, n$). The latter case may arise on account of the learning curve effect associated with fixed ordering costs over time. Moreover, when all the future costs are calculated in terms of their present values, even if the undiscounted fixed cost may increase over time, Condition (2.18) still holds as long as the rate of increase of the fixed cost over time is less than or equal to the discount rate.

Theorems C.2.2 and C.2.3 are included in Appendix C to provide the required existing results on K-convex functions or their extensions. We can now derive the following result.

THEOREM 2.3 *Assume* (2.18) *in addition to the assumptions made in Section* 2.3. *Then there exists a sequence of numbers* $s_{n,i}, S_{n,i}$, $n \in \langle 0, N-1 \rangle, i \in \mathbb{I}$, *with* $s_{n,i} \leq S_{n,i}$, *such that the optimal feedback policy is*

$$\hat{u}_n(i,x) = (S_{n,i} - x)\mathbb{1}_{x<s_{n,i}}. \tag{2.19}$$

Proof. The dynamic programming equations (2.12) can be written as

$$\begin{cases} v_n(i,x) = f_n(i,x) - c_n^i x + h_n(i,x), & n \in \langle 0, N-1 \rangle,\ i \in \mathbb{I}, \\ v_N(i,x) = f_N(i,x), & i \in \mathbb{I}, \end{cases} \tag{2.20}$$

where

$$h_n(i,x) = \inf_{y \geq x} [K_n^i \mathbb{1}_{y>x} + z_n(i,y)] \tag{2.21}$$

and

$$z_n(i,y) = c_n^i y + F_{n+1}(v_{n+1})(i,y). \tag{2.22}$$

From (2.5) and (2.12), we have $v_n(i,x) \geq f_n(i,x)$, $\forall n \in \langle 0, N-1 \rangle$. From Theorem 2.1, we know that $v_n \in \mathbb{C}_1$. These, along with (2.11), ensure for $n \in \langle 0, N-1 \rangle$ and $i \in \mathbb{I}$, that $z_n(i,y) \to +\infty$ as $y \to \infty$, and $z_n(i,y)$ is uniformly continuous.

In order to apply Theorem C.2.3 to obtain (2.19), we need only to prove that $z_n(i,x)$ is K_n^i-convex. According to Theorem C.2.2, it is sufficient to show that $v_{n+1}(i,x)$ is K_{n+1}^i-convex. This is done by induction. First, $v_N(i,x)$ is convex by definition and, therefore, K-convex for any $K \geq 0$. Let us now assume that for a given $n \leq N-1$ and i, $v_{n+1}(i,x)$ is K_{n+1}^i-convex. By Theorem C.2.2 and Condition (2.18), it is easy to see that $z_n(i,x)$ is $\bar{K}_{n+1}^i$-convex, hence also K_n^i-convex. Then, Theorem C.2.3 implies that $h_n(i,x)$ is K_n^i-convex. Therefore, $v_n(i,x)$ is K_n^i-convex. This completes the induction argument.

Thus, it follows that $z_n(i,x)$ is K_n^i-convex for each n and i. Since $z_n(i,y) \to +\infty$ when $y \to \infty$, we apply Theorem C.2.3 to obtain the desired $s_{n,i}$ and $S_{n,i}$. According to Theorem 2.2, the (s,S)-type policy defined in (2.19) is optimal. □

REMARK 2.3 Theorem 2.3 can be easily extended to allow for a constant leadtime in the delivery of orders. The usual approach is to replace the surplus level by the so-called *surplus position.* It can also be generalized to Markovian demands with discrete components and countably many states.

REMARK 2.4 In the standard model with $L = 1$, Veinott (1966) gives an alternate proof to the one by Scarf (1960) based on K-convexity. For this, he does not need a condition like (2.11), but requires other assumptions instead.

2.6. Nonstationary Infinite Horizon Problem

We now consider an infinite horizon version of the problem formulated in Section 2.3. By letting $N = \infty$ and $U = (u_n, u_{n+1}, \ldots)$, the extended real-valued objective function of the problem becomes

$$J_n(i, x; U) = \sum_{k=n}^{\infty} \alpha^{k-n} \mathsf{E}[c_k(i_k, u_k) + f_k(i_k, x_k)], \tag{2.23}$$

where α is a given discount factor, $0 < \alpha \leq 1$. The dynamic programming equations are

$$v_n(i, x) = f_n(i, x) + \inf_{u \geq 0} \{c_n(i, u) + \alpha F_{n+1}(v_{n+1})(i, x+u)\},\ n = 0, 1, 2, \ldots. \tag{2.24}$$

In what follows, we will show that there exists a solution of (2.24) in class $\mathbb{C}_1$, which is the value function of the infinite horizon problem; (see also Remark 2.5). Moreover, the decision that attains the infimum in (2.24) is an optimal feedback policy. Our method is that of successive approximation of the infinite horizon problem by longer and longer finite horizon problems.

Therefore, we examine the finite horizon approximation $J_{n,m}(i, x; U)$, $m \geq 1$, of (2.23), which is obtained by the first m-period truncation of the infinite horizon problem of minimizing $J_n(i, x; U)$, i.e.,

$$J_{n,m}(i, x; U) = \sum_{k=n}^{n+m-1} \alpha^{k-n} \mathsf{E}[c_k(i_k, u_k) + f_k(i_k, x_k)]. \tag{2.25}$$

Let $v_{n,m}(i, x)$ be the value function of the truncated problem, i.e.,

$$v_{n,m}(i, x) = \inf_{U \in \mathcal{U}} J_{n,m}(i, x; U). \tag{2.26}$$

Since (2.26) is a finite horizon problem on the interval $\langle n, n+m \rangle$, we may apply Theorems 2.1 and 2.2 and obtain its value function by solving the dynamic programming equations

$$\begin{cases} v_{n,m+1}(i, x) &= f_n(i, x) + \inf_{u \geq 0} \{c_n(i, u) + \alpha F_{n+1}(v_{n+1,m})(i, x+u)\}, \\ v_{n+m,0}(i, x) &= 0. \end{cases} \tag{2.27}$$

Moreover, $v_{n,0}(i,x) = 0$, $v_{n,m} \in \mathbb{C}_1$, and the infimum in (2.26) is attained.

It is not difficult to see that the value function $v_{n,m}$ increases in m. In order to take its limit as $m \to \infty$, we need to establish an upper bound on $v_{n,m}$. One possible upper bound on $\inf_{U \in \mathcal{U}} J_n(i,x;u)$ can be obtained by computing the objective function value associated with a policy of never ordering anything. With the notation $\mathbf{0} = \{0, 0, \ldots\}$ for this policy, let us write

$$\begin{aligned} w_n(i,x) &= J_n(i,x;\mathbf{0}) \\ &= f_n(i,x) + \mathsf{E}\left\{ \sum_{k=n+1}^{\infty} \alpha^{k-n} f_k(i_k, x - \sum_{j=1}^{k-1} \xi_j) | i_n = i \right\}. \end{aligned} \tag{2.28}$$

In a way similar to Section I.5.1 of Chapter 4 in Bensoussan *et al.* (1983), it is easy to see that given (2.6), $w_n(i,x)$ is well-defined and is in $\mathbb{C}_1$. Furthermore, in class $\mathbb{C}_1$, w_n is the unique solution of

$$w_n(i,x) = f_n(i,x) + \alpha F_{n+1}(w_{n+1})(i,x). \tag{2.29}$$

We can state the following result for the infinite horizon problem.

THEOREM 2.4 *Assume* (2.5) *and* (2.6). *Then, we have*

$$0 = v_{n,0} \le v_{n,1} \le \ldots \le v_{n,m} \le w_n \tag{2.30}$$

and as $m \to \infty$

$$v_{n,m} \uparrow v_n, \text{ a solution of (2.24) in } \mathbb{B}_1. \tag{2.31}$$

Furthermore, $v_n \in \mathbb{C}_1$, *and we can obtain* $\hat{U} = \{\hat{u}_n, \hat{u}_{n+1}, \ldots\}$ *for which the infimum in* (2.24) *is attained. Moreover,* $\hat{U}$ *is an optimal feedback policy, i.e.,*

$$v_n(i,x) = \min_{U \in \mathcal{U}} J_n(i,x;U) = J_n(i,x;\hat{U}). \tag{2.32}$$

Proof. By definition, $v_{n,0} = 0$. Let $\tilde{U}_{n,m} = \{\tilde{u}_n, \tilde{u}_{n+1}, \ldots, \tilde{u}_{n+m-1}\}$ be a minimizer of (2.25). Thus,

$$\begin{aligned} v_{n,m}(i,x) &= J_{n,m}(i,x;\tilde{U}_{n,m}) \ge J_{n,m-1}(i,x;\tilde{U}_{n,m}) \\ &\ge \min_{U \in \mathcal{U}} J_{n,m-1}(i,x;U) = v_{n,m-1}(i,x). \end{aligned}$$

It is also obvious from (2.25) and (2.28) that $v_{n,m}(i,x) \le J_{n,m}(i,x;\mathbf{0}) \le w_n(i,x)$. This proves (2.30). Since $v_{n,m} \in \mathbb{C}_1$, we have

$$v_{n,m}(i,x) \uparrow v_n(i,x) \le w_n(i,x), \tag{2.33}$$

with $v_n(i,x)$ l.s.c., and hence in $\mathbb{B}_1$. Next, we show that v_n satisfies the dynamic programming equations (2.24). Observe from (2.27) and (2.30) that for each m, we have

$$v_{n,m}(i,x) \leq f_n(i,x) + \inf_{u\geq 0}\{c_n(i,u) + \alpha F_{n+1}(v_{n+1,m})(i,x+u)\}.$$

Thus, in view of (2.33), we obtain

$$v_n(i,x) \leq f_n(i,x) + \inf_{u\geq 0}\{c_n(i,u) + \alpha F_{n+1}(v_{n+1})(i,x+u)\}. \tag{2.34}$$

In order to obtain the reverse inequality, let $\hat{u}_{n,m}$ attain the infimum on the RHS of (2.27). From (2.5) and (2.6), we obtain that

$$c_n^i \hat{u}_{n,m}(i,x) \leq \alpha F_{n+1}(v_{n+1,m})(i,x) \leq \alpha(1+M) \ \| w_n \| \ (1+|x|),$$

where $\|\cdot\|$ is the norm defined on $\mathbb{B}_1$. This provides us with the bound

$$0 \leq \hat{u}_{n,m}(i,x) \leq M_n(1+|x|). \tag{2.35}$$

For $l > m$, we see from (2.27) that

$$\begin{aligned} v_{n,l+1}(i,x) &= f_n(i,x) + c_n(i,\hat{u}_{n,l}(x)) \\ &\quad +\alpha F_{n+1}(v_{n+1,l})(i,x+\hat{u}_{n,l}(x)) \\ &\geq f_n(i,x) + c_n(i,\hat{u}_{n,l}(x)) \\ &\quad +\alpha F_{n+1}(v_{n+1,m})(i,x+\hat{u}_{n,l}(x)). \end{aligned} \tag{2.36}$$

Fix m and let $l \to \infty$. In view of (2.35), we can, for any given n, i and x, extract a subsequence $\hat{u}_{n,l'}(i,x)$ such that $\hat{u}_{n,l'}(i,x) \to \bar{u}_n(i,x)$. Since $v_{n+1,m}$ is uniformly continuous in m and c_n is l.s.c., we can pass to the limit on the RHS of (2.36). Noting that the left-hand side converges as well, we obtain

$$\begin{aligned} v_n(i,x) &\geq f_n(i,x) + c_n(i,\bar{u}_n(x)) + \alpha F_{n+1}(v_{n+1,m})(i,x+\bar{u}_n(x)) \\ &\geq f_n(i,x) + \inf_{u\geq 0}\{c_n(i,u) + \alpha F_{n+1}(v_{n+1,m})(i,x+u)\}. \end{aligned}$$

This, along with (2.33), (2.34), and the fact that $v_n(i,x) \in \mathbb{B}_1$, proves (2.31).

Next we prove that $v_n \in \mathbb{C}_1$. Let us consider the problem (2.25) again. From (2.23),

$$J_n(i,x';U) - J_n(i,x;U)$$

$$= \sum_{k=n}^{\infty} \alpha^{k-n} \mathsf{E} \left[f_k \left(i_k, x' + \sum_{j=n}^{k-1} u_j - \sum_{j=n+1}^{k} \xi_i \right) \right.$$
$$\left. - f_k \left(i_k, x + \sum_{j=n}^{k-1} u_j - \sum_{j=n+1}^{k} \xi_i \right) \right].$$

From (2.6), we have

$$|J_{n,m}(i,x';U) - J_{n,m}(i,x;U)| \le \sum_{l=n}^{\infty} \alpha^{l-n} C|x'-x| = C|x'-x|/(1-\alpha),$$

which implies $|v_{n,m}(i,x') - v_{n,m}(i,x)| \le C|x'-x|/(1-\alpha)$. By taking the limit as $m \to \infty$, we have $|v_n(i,x') - v_n(i,x)| \le C|x'-x|/(1-\alpha)$, from which it follows that $v_n \in \mathbb{C}_1$. Therefore, there exists a function $\hat{u}_n(i,x)$ in $\mathbb{B}_0$ such that

$$\begin{aligned} c_n(i, \hat{u}_n(i,x)) \quad &+ \quad \alpha F_{n+1}(v_{n+1})(i, x + \hat{u}_n(i,x)) \\ &= \quad \inf_{u \ge 0} \{ c_n(i,u) + \alpha F_{n+1}(v_{n+1})(i, x+u) \}. \end{aligned}$$

Hence, we have

$$v_n(i,x) = J_n(i,x;\hat{U}) \ge \inf_{U \in \mathcal{U}} J_n(i,x;U).$$

But for any arbitrary admissible control U, we also know that $v_n(i,x) \le J_n(i,x;U)$. Therefore, we conclude that

$$v_n(i,x) = J_n(i,x;\hat{U}) = \min_{U \in \mathcal{U}} J_n(i,x;U).$$

□

REMARK 2.5 We should indicate that Theorem 2.4 does not imply that there is a unique solution of the dynamic programming equations (2.24). There may well be other solutions. Moreover, one can show that the value function is the minimal positive solution of (2.24). It is also possible to obtain a uniqueness proof under additional assumptions.

With Theorem 2.4 in hand, we can now prove the optimality of an (s,S)-type policy for the nonstationary infinite horizon problem.

THEOREM 2.5 *Assume* (2.5), (2.6), *and* (2.11) *hold for the infinite horizon problem. Then, there exists a sequence of numbers* $s_{n,i}, S_{n,i},\ n =$

$0, 1, \ldots,$ *with* $s_{n,i} \leq S_{n,i}$ *for each* $i \in \mathbb{I}$, *such that the optimal feedback policy is* $\hat{u}_n(i, x) = (S_{n,i} - x)\mathbb{1}_{x<s_{n,i}}$.

Proof. Let v_n denote the value function. Define the functions z_n and h_n as in Section 2.5. We know that $z_n(i, x) \to \infty$ as $x \to +\infty$ and $z_n(i, x) \in \mathbb{C}_1$ for all n and $i \in \mathbb{I}$.

We now prove that v_n is K_n^i-convex. Using the same induction as in Section 2.5, we can show that $v_{n,k}(i, x)$, as defined in (2.26), is K_n^i-convex. This induction is possible since we know that $v_{n,k}(i, x)$ satisfies the dynamic programming equations (2.27). It is clear from the definition of K-convexity and from taking the limit as $k \to \infty$, that the value function $v_n(i, x)$ is also K_n^i-convex.

From Theorem 2.4, we know that $v_n \in \mathbb{C}_1$ and that v_n satisfies the dynamic programming equations (2.24). Therefore, we can obtain an optimal feedback policy $\hat{U} = \{\hat{u}_n, \hat{u}_{n+1}, \ldots\}$ that attains the infimum in (2.24). Because v_n is K_n^i-convex, $\hat{u}_n$ can be expressed as in Theorem 2.5. □

2.7. Cyclic Demand Model

Cyclic or seasonal demand often arises in practice. Such a demand represents a special case of the Markovian demand, where the number of demand states L is given by the cycle length, and

$$p_{ij} = \begin{cases} 1, & \text{if } j = i + 1, i = 1, \ldots, L - 1, \text{ or } i = L, j = 1, \\ 0, & \text{otherwise.} \end{cases}$$

Furthermore, we assume that the cost functions and density functions are all time invariant. The result is a considerably simplified optimal policy, i.e., only L pairs of (s_n, S_n) need to be computed.

We can state the following corollary to Theorem 2.5.

COROLLARY 2.1 *In the infinite horizon inventory problem with the demand cycle of* L *periods, let* n_1 *and* n_2 $(n_1 < n_2)$ *be any two periods such that* $n_2 = n_1 + m \cdot L$, $m = 1, 2, \ldots$. *Then, we have* $s_{n_1} = s_{n_2}$ *and* $S_{n_1} = S_{n_2}$.

2.8. Constrained Models

In this section, we incorporate some additional constraints that often arise in practice. We show that (s, S)-type policies continue to remain optimal for the extended models.

2.8.1 No-ordering Periods

Consider the special situation in which ordering is not possible in certain periods (for example, suppliers do not accept orders on weekends). We will show that the following theorem holds in such a situation.

THEOREM 2.6 *In the problem with some no-ordering periods, the optimal policy is still of (s, S)-type for any period, except when the ordering is not allowed.*

Proof. To stay with our earlier notation, it is no loss of generality to continue assuming the setup cost to be K_m^i in a no-ordering period m with the demand state i; clearly, setup costs are of no use in no-ordering periods. The definition (2.21) is revised as

$$h_n(i,x) = \begin{cases} z_n(i,x), & \text{in a no-ordering period } n, \\ \inf\limits_{y \geq x}[K_n^i \mathbb{1}_{y>x} + z_n(i,y)], & \text{otherwise,} \end{cases} \tag{2.37}$$

and $z_n(i,y)$ is defined as before in (2.22). Using the same induction argument as in the proof of Theorem 2.3, we can show that $h_n(i,x)$ and $v_n(i,x)$ are K_n^i-convex if ordering is allowed in period n. If ordering is disallowed in period n, then $h_n(i,x) = z_n(i,x)$, which is $\bar{K}_{n+1}^i$-convex, and therefore also K_n^i-convex. Thus, in both cases $v_n(i,x)$ is K_n^i-convex. □

REMARK 2.6 Theorem 2.6 can be generalized to allow for supply uncertainty as in Parlar *et al.* (1995). One needs to replace u_k in (2.8) by $a_k u_k$, where $\mathsf{P}\{a_k = 1 | i_k = i\} = q_k^i$ and $\mathsf{P}\{a_k = 0 | i_k = i\} = 1 - q_k^i$, and modify (3.2) appropriately.

2.8.2 Storage and Service Level Constraints

Let $B < \infty$ denote an upper bound on the inventory level. Moreover, to guarantee a reasonable measure of service level, we introduce a chance constraint requiring that the probability of the ending inventory falling below zero in any given period does not exceed $1 - \alpha_p$ for a specified $\alpha_p \in (0, 1]$, known as Type 1 service level. Thus,

$$\mathsf{P}\{x_{k+1} < 0\} \leq 1 - \alpha_p, \; k \in \langle 0, N-1 \rangle.$$

As an example, if we set $\alpha_p = 0.95$, then we are requiring that we satisfy the demand in any given period with at least 95% probability.

Given the demand state i in period k and the inventory dynamics (2.8), we can write the above condition as $\Phi_{i,k}(x_k + u_k) \geq \alpha_p$, which can be converted into $x_k + u_k \geq A_k^i$, where $A_k^i = \inf(a | \Phi_{i,k}(a) \geq \alpha_p)$, referred to as the safety stock in period k that guarantees the Type 1 service level α_p in that period. If we define the *quantile function* $\Phi_{i,k}^{-1}(\eta) = \inf(a|\Phi_{i,k}(a) \geq \eta)$, then we can also write $A_k^i = \Phi_{i,k}^{-1}(\alpha_p)$.

The dynamic programming equations can be written as (2.22), where $z_n(i,y)$ is as in (2.22) and

$$h_n(i,x) = \inf_{y \geq x, A_n^i \leq y \leq B} [K_n^i \mathbb{I}_{y>x} + z_n(i,y)], \tag{2.38}$$

provided $A_n^i \leq B$, $n \in \langle 0, N-1 \rangle, i \in \mathbb{I}$; if not, then there is no feasible solution, and $h_n(i,x) = \inf \emptyset \equiv \infty$. This time, since y is bounded by $B < \infty$, Theorem 2.1 can be relaxed as follows.

THEOREM 2.7 *The dynamic programming equations* (2.20) *with* (2.38) *define a sequence of l.s.c. functions on* $(-\infty, B]$. *Moreover, there exists a function* $\hat{u}_n(i,x)$ *in* $\mathbb{B}_0$, *which attains the infimum in* (2.20) *for any* $x \in (-\infty, B]$.

With $\hat{u}_n(i,x)$ of Theorem 2.7, it is possible to prove Theorem 2.2, also known as the verification theorem, for the constrained case. We now show that the optimal policy is of (s,S)-type.

THEOREM 2.8 *There is a sequence of numbers* $s_{n,i}$, $S_{n,i}$, $n \in \langle 0, N-1 \rangle$, $i \in \mathbb{I}$, *with* $s_{n,i} \leq S_{n,i}$ *and* $A_n^i \leq S_{n,i} \leq B$ *such that optimal feedback policy* $\hat{u}_n(i,x) = (S_{n,i} - x)\mathbb{I}_{x<s_{n,i}}$ *is optimal for the model with capacity and service constraints defined above.*

Proof. First note that Theorem C.2.3 holds when g is l.s.c. and K-convex on $(-\infty, B]$, $B < \infty$. Also, by Theorem C.2.2 (iii) and (iv), one can see that $\mathsf{E}g(x-\xi)$ is K-convex on $(-\infty, B]$ since $\xi \geq 0$. Because g is l.s.c., it is easily seen that $\mathsf{E}g(x-\xi)$ is l.s.c. on $(-\infty, B]$. Furthermore, by Theorem 2.7, v_n is l.s.c. on $(-\infty, B]$. With these observations in mind, the proof of Theorem 2.3 can easily be modified to complete the proof. □

REMARK 2.7 A constant integer leadtime $\tau \geq 1$ can also be included in this model, with the surplus level replaced by the surplus position and with the lower bound A_k^i properly redefined in terms of the distribution of the total demand during the leadtime; (see, e.g., Porteus (1971) or Zipkin (2000)).

2.9. Concluding Remarks and Notes

This chapter, based on Sethi and Cheng (1997), develops various realistic extensions of the classical dynamic inventory model with stochastic demands. The models consider demands that are dependent on a finite state Markov chain including demands that are cyclic. Some constraints commonly encountered in practice, namely no-ordering periods, finite storage capacities, and service levels, are also treated. Both finite and infinite horizon cases are studied. It is shown that all of these models, not unlike the classical model, exhibit the optimality of (s, S)-type policies.

Chapter 3

DISCOUNT COST MODELS WITH POLYNOMIALLY GROWING SURPLUS COST

3.1. Introduction

This chapter studies stochastic inventory problems with unbounded Markovian demands and more general costs than those considered in Chapter 2. Finite horizon problems, as well as stationary and nonstationary discounted cost infinite horizon problems, are addressed. Existence of optimal Markov or feedback policies is established with Markovian demand: unbounded, ordering costs that are l.s.c., and surplus costs that are l.s.c. with polynomial growth. Furthermore, optimality of (s, S)-type policies is proved when the ordering cost consists of fixed and proportional cost components and the surplus cost is convex.

The literature on infinite horizon inventory models involving a fixed ordering cost assumes surplus cost to be of linear growth and uniformly continuous as in Karlin (1958c), Scarf (1960), Bensoussan *et al.* (1983), and others. Even quadratic surplus costs that are popular in the production planning literature dating back to the classical HMMS model of Holt *et al.* (1960) have not been considered in infinite horizon inventory models.

As for demand, Karlin and Fabens (1960), Song and Zipkin (1993), Sethi and Cheng (1997), and Beyer and Sethi (1997) have all considered Markovian demands. Karlin and Fabens consider only the class of state-independent (s, S) policies, which does not in general include optimal policies. Song and Zipkin consider Markov-modulated Poisson demand in their analysis. Sethi and Cheng consider general Markovian demands in their treatment of discounted cost problems.

In this chapter, we consider unbounded Markovian demands but require that a certain number (depending on the growth rate of the surplus

D. Beyer et al., *Markovian Demand Inventory Models*, International Series in Operations Research & Management Science 108,
DOI 10.1007/978-0-387-71604-6_3, © Springer Science+Business Media, LLC 2010

cost function) of moments be finite. This is an essential requirement and yet not very restrictive. As both cost and demand are generalized, the chapter represents a significant extension of the infinite horizon inventory problems (involving a fixed ordering cost component) that have appeared in the literature.

In this chapter, we conduct a detailed analysis of the discounted cost problem. The problem is carefully formulated in Section 3.2. In Section 3.3, we use the dynamic programming equation to prove the existence of an optimal Markov control for the finite horizon problem. We also provide a verification theorem for the solution of the dynamic programming equation to be the value function. We prove the value function to be continuous when the surplus cost is continuous and the ordering cost is l.s.c. As we will remark later, this has some implications for *whether or not to order at the level* s in an optimal (s, S)-type policy. The nonstationary infinite horizon problem is treated in Section 3.4. With further assumptions on costs, the optimality of (s, S)-type policies is established in Section 3.5. The stationary infinite horizon problem is briefly discussed in Section 3.6. The chapter concludes with end notes in Section 3.7.

3.2. Formulation of the Model

Let us consider an inventory problem over a finite number of periods $\langle n, N\rangle = \{n, n+1, \ldots, N\}$, and an initial inventory of x units at the beginning of period n, where n and N are any given integers satisfying $0 \le n \le N < \infty$. The demand in each period is assumed to be a random variable defined on a given probability space $(\Omega, \mathcal{F}, \mathsf{P})$, and not necessarily identically distributed. More specifically, the demand distributions in successive periods are defined as below.

Consider a finite collection of demand states $\mathbb{I} = \{1, 2, \ldots, L\}$, and let i_k denote the demand state in the kth period. We assume that i_k, $k \in \langle n, N\rangle$, with known initial demand state i_n, is a Markov chain over $\mathbb{I}$ with the transition matrix $P = \{p_{ij}\}$. Thus,

$$0 \le p_{ij} \le 1,\ i \in \mathbb{I}, j \in \mathbb{I}, \ \text{ and } \sum_{j=1}^{L} p_{ij} = 1,\ i \in \mathbb{I}.$$

Let a nonnegative random variable ξ_k denote the demand in a given period k, $k = 0, \ldots, N-1$. Demand ξ_k depends only on period k and the demand state in that period, by which we mean that it is independent of past demand states and past demands. We denote its cumulative probability distribution by $\Phi_{i,k}(x)$, when the demand state $i_k = i$. In the following period, if the state changes to state j, which happens with probability p_{ij}, then the demand distribution is $\Phi_{j,k+1}$ in that period. We further assume that for some $\gamma \geq 1$ and a positive constant D,

$$\mathsf{E}(\xi_k^\gamma | i_k = i) = \int_0^\infty x^\gamma d\Phi_{i,k}(x) \leq D < \infty, \; k = 0, \ldots, N-1, \; i \in \mathbb{I}. \quad (3.1)$$

This is not a very restrictive assumption from an applied perspective.

We denote by

$$\begin{aligned} &\mathcal{F}_l^k, \text{ the } \sigma\text{-algebra generated by } \{i_l, \ldots, i_{k-1}, i_k; \xi_l, \ldots, \xi_{k-1}\}, \\ &\quad 0 \leq l \leq k \leq N, \\ &\mathcal{F}^k = \mathcal{F}_0^k. \end{aligned} \quad (3.2)$$

Since i_k, $k = 1, \ldots, N$, is a Markov chain and ξ_k depends only on i_k, we have

$$\begin{aligned} \mathsf{E}(\xi_k | \mathcal{F}^k) &= \mathsf{E}(\xi_k | i_0, i_1, \ldots, i_k; \xi_0, \xi_1, \ldots, \xi_{k-1}) \\ &= \mathsf{E}(\xi_k | i_k). \end{aligned} \quad (3.3)$$

An admissible decision (ordering quantities) for the problem on the interval $\langle n, N \rangle$ with initial state $i_n = i$ can be denoted as

$$U = (u_n, \ldots, u_{N-1}), \quad (3.4)$$

where u_k is a nonnegative $\mathcal{F}_n^k$-measurable random variable. In simpler terms, this means that decision u_k depends only on the past information. Note that since i_n is known in period n, $\mathcal{F}_n^n = (\Omega, \emptyset)$; hence u_n is deterministic. Moreover, it should be emphasized that this class of admissible decisions is larger than the class of admissible feedback policies.

Ordering quantities are decided upon at the beginning of each period. Demand in each period is supposed to occur at the end of the period after the order has been delivered. Unsatisfied demand is carried forward as backlog. The inventory balance equations are defined by

$$\begin{cases} x_{k+1} = x_k + u_k - \xi_k, \; k = n, \ldots, N-1, \\ x_n = x, \; \text{ initial inventory level}, \\ i_k, k = n, \ldots, N, \text{ Markov chain with transition matrix } P, \\ i_n = i, \; \text{ initial state}, \end{cases}$$

where x_k is the surplus level at the beginning of period k, u_k is the quantity ordered at the beginning of period k, i_k is the demand state in period k, and ξ_k is the demand in period k. Note that $x_k > 0$ represents an inventory of x_k and $x_k < 0$ represents a backlog (or shortage) of $-x_k$.

Next, we specify the relevant costs and the assumptions they satisfy.

(i) The production cost function $c_k(i,u) : \mathbb{I} \times \mathbb{R}^+ \to \mathbb{R}^+$ is l.s.c., $c_k(i,0) = 0$, $k = 0,1,\ldots,N-1$.

(ii) The surplus cost function $f_k(i,x) : \mathbb{I} \times \mathbb{R}^+ \to \mathbb{R}^+$ is l.s.c., with $f_k(i,x) \leq \bar{f}(1+|x|^\gamma)$, $i = 1,2\ldots,L$, $k = 0,1,\ldots,N-1$, where $\bar{f}$ is a nonnegative constant. When $x < 0$, $f_k(i,x)$ is the cost of backlogged sales x, and when $x > 0$, $f_k(i,x)$ is the carrying cost of holding inventory x during that period.

(iii) $f_N(i,x) : \mathbb{I} \times \mathbb{R}^+ \to \mathbb{R}^+$, the penalty cost/disposal cost for the terminal surplus, is l.s.c. with $f_N(i,x) \leq \bar{f}(1+|x|^\gamma)$. When $x < 0$, $f_N(i,x)$ represents the penalty cost of unsatisfied demand x, and when $x > 0$, $f_N(i,x)$ represents the disposal cost of the inventory level x.

The objective function to be minimized is the expected present value of all the costs incurred during the interval $\langle n, N\rangle$, i.e.,

$$\begin{aligned} J_n(i,x;U) &= \mathsf{E}\left\{\sum_{k=n}^{N-1} \alpha^{k-n}[c_k(i_k,u_k) + f_k(i_k,x_k)]\right. \\ &\quad \left. +\alpha^{N-n} f_N(i_N,x_N)\right\}, \end{aligned} \tag{3.5}$$

which is always defined, provided we allow this quantity to be infinite.

In (3.5), α denotes the discount factor with $0 < \alpha < 1$. While introducing α in this finite horizon model does not add to generality, in view of the already time-dependent nature of the cost functions, we do so for convenience of exposition in dealing with the average cost optimality criterion studied in Chapter 6.

3.3. Dynamic Programming and Optimal Feedback Policy

Let us first introduce the following definitions.

$\mathbb{B}^\gamma$= Banach space of Borel functions $b : \mathbb{I} \times \mathbb{R} \to \mathbb{R}$ with polynomial growth with power γ or less. More specifically, if $b \in \mathbb{B}^\gamma$, then $|b(i,x)| \leq \| b \|_\gamma (1+|x|^\gamma)$, where the norm

$$\| b \|_\gamma = \max_i \sup_x \frac{|b(i,x)|}{1+|x|^\gamma} < \infty. \tag{3.6}$$

$\mathbb{L}^\gamma$ = the subspace of l.s.c. functions in $\mathbb{B}^\gamma$. The space $\mathbb{L}^\gamma$ is closed in $\mathbb{B}^\gamma$.

$\mathbb{L}^{\gamma^-}$ = the class of all l.s.c. functions which are of polynomial growth with power γ or less on $(-\infty, 0]$.

In view of (3.3), we can write

$$\mathsf{E}[b(i_{k+1}, \xi_k)|\mathcal{F}^k] = \mathsf{E}[b(i_{k+1}, \xi_k)|i_k], \tag{3.7}$$

for any $b \in \mathbb{B}^\gamma$.

To write the dynamic programming equations more concisely, we define the operator F_{n+1} on $\mathbb{B}^\gamma$ as follows:

$$\begin{aligned} F_{k+1}b(i,y) &= \mathsf{E}[b(i_{k+1}, y-\xi_k)|i_k = i] \\ &= \sum_{j=1}^{L} \{\mathsf{P}(i_{k+1} = j|i_k = i)\mathsf{E}[b(j, y-\xi_k)|i_k = i]\} \\ &= \sum_{j=1}^{L} p_{ij} \int_0^\infty b(j, y-\xi) d\Phi_{i,k}(\xi). \end{aligned} \tag{3.8}$$

In addition to Assumptions (i)-(iii) on costs, we also require that for $k = 0, 1, \ldots, N-1$,

$$0 \le c_k(i,u) + \alpha F_{k+1}(f_{k+1})(i,u) \to \infty \text{ for } u \to \infty. \tag{3.9}$$

REMARK 3.1 Condition (3.9) implies that both the purchase cost and the inventory (or salvage) cost associated with a decision in any given period cannot both be zero. The conditions rule out the trivial and unrealistic situation of ordering an infinite amount as the optimal policy. See Remark 2.2 for further elaboration.

Let $v_n(i,x)$ represent the optimal value of the expected costs during the time horizon $\langle n, N\rangle$ with demand state i in period n, i.e.,

$$v_n(i,x) = \inf_U J_n(i,x;U).$$

Then, $v_n(i,x)$ satisfies the dynamic programming equations

$$\begin{aligned} v_n(i,x) &= f_n(i,x) + \inf_{u \ge 0} \{c_n(i,u) + \alpha \mathsf{E}[v_{n+1}(i_{n+1}, \\ &\qquad x+u-\xi_n)|i_n = i]\} \\ &= f_n(i,x) + \inf_{u\ge 0}\{c_n(i,u) + \alpha F_{n+1}(v_{n+1})(i, x+u)\}, \\ &\qquad n = 0, 1, \ldots, N-1, \end{aligned} \tag{3.10}$$

$$v_N(i,x) = f_N(i,x). \tag{3.11}$$

We can now state our existence results in the following two theorems.

THEOREM 3.1 *The dynamic programming equations* (3.10) *and* (3.11) *define a sequence of functions in* $\mathbb{L}^\gamma$. *Moreover, there exists a Borel function* $\hat{u}_n(i,x)$ *such that the infimum in* (3.10) *is attained at* $u = \hat{u}_n(i,x)$ *for any* x. *Furthermore, if the functions* $f_n(i,\cdot)$, $n = 0,1,\ldots,N$, $i = 1,2\ldots,L$, *are continuous, then the functions defined by* (3.10) *and* (3.11) *are continuous.*

Proof. We proceed by induction. Because of Assumption (iii) and (3.11), it follows that $v_N(i,x) \in \mathbb{L}^\gamma$. Assume $v_{n+1}(i,x)$ belongs to $\mathbb{L}^\gamma$. Consider points x such that $|x| \leq M$, where M is an arbitrary nonnegative integer. Let

$$B_{n,i}^M = \sup_{|x|\leq M} \{\alpha F_{n+1}(v_{n+1})(i,x)\}. \tag{3.12}$$

The constant $B_{n,i}^M$ is finite since $v_{n+1}(i,x)$ is in $\mathbb{B}^\gamma$ and therefore bounded on $|x| \leq M$, and F_{n+1} is a continuous linear operator; (see Lemma A.4.1). Because of (3.9), we know that the set

$$N_{n,i}^M := \{u \geq 0 : \inf_{|x|\leq M} \{c_n(i,u) + \alpha F_{n+1}(f_{n+1})(i,x+u) \leq B_{n,i}^M\} \tag{3.13}$$

is bounded, i.e., there is a $\bar{u}_{n,i}^M$ such that

$$N_{n,i}^M \subseteq [0, \bar{u}_{n,i}^M]. \tag{3.14}$$

Because of $v_{n+1} \geq f_{n+1}$, we conclude that

$$\begin{aligned}\{u \geq 0 : \inf_{|x|\leq M} \{c_n(i,u) + \alpha F_{n+1}(v_{n+1})(i,x+u) \leq B_{n,i}^M\} \\ \subseteq N_{n,i}^M \subseteq [0, \bar{u}_{n,i}^M],\end{aligned} \tag{3.15}$$

and, therefore without loss of optimality, we can restrict our attention to $0 \leq u \leq \bar{u}_{n,i}^M$ for all x satisfying $|x| \leq M$. This is because for any $u > \bar{u}_{n,i}^M$,

$$\begin{aligned} c_n(i,u) &+ \alpha F_{n+1}(v_{n+1})(i,x+u) \\ &\geq \inf_{|x|\leq M} \{c_n(i,u) + \alpha F_{n+1}(v_{n+1})(i,x+u)\} \\ &> B_{n,i}^M = \sup_{|x|\leq M} \{\alpha F_{n+1}(v_{n+1})(i,x)\} \\ &\geq \alpha F_{n+1}(v_{n+1})(i,x) \\ &= c_n(i,0) + \alpha F_{n+1}(v_{n+1})(i,x), \end{aligned}$$

and thus u cannot be the point where the infimum is attained.

Since the function

$$\psi_n(i,x,u) = c_n(i,u) + \alpha F_{n+1}(v_{n+1})(i,x+u)$$

is l.s.c. and bounded from below, its minimum over a compact set is attained; (see Theorem A.1.5). Moreover, from the classical Selection Theorem A.1.7, we know that there exists a Borel function $\hat{u}_n^M(i,x)$ such that

$$\psi_n(i,x,\hat{u}_n^M(i,x)) = \inf_{0\le u\le \bar{u}_{n,i}^M} \psi_n(i,x,u), \ |x| \le M. \tag{3.16}$$

Upon defining

$$\hat{u}_n(i,x) = \hat{u}_n^M(i,x) \ \text{ for } M-1 < |x| \le M,$$

we obtain a Borel function such that

$$\psi_n(i,x,\hat{u}_n(i,x)) = \inf_{u\ge 0} \psi_n(i,x,u), \ \forall x. \tag{3.17}$$

Since

$$\inf_{u\ge 0} \psi_n(i,x,u) \le \psi_n(i,x,0) \le c_n(i,0) + \alpha \parallel F_{n+1} \parallel \parallel v_{n+1} \parallel_\gamma (1+|x|^\gamma),$$

we can use (3.10), Assumption (ii), and Lemma A.4.1 to conclude that $v_n(i,x) \in \mathbb{B}^\gamma$. Furthermore, because $\psi_n(i,\cdot,\cdot)$ is l.s.c., it follows from equation (3.16) that $v_n(i,\cdot)$ is l.s.c. for each i (see Theorem A.1.6), and therefore $v_n \in \mathbb{L}^\gamma$.

To prove the last part of the theorem, we begin with the fact that for each i the function $f_n(i,\cdot)$ is continuous, $n = 0,1,\ldots,N$. Then $v_N(i,\cdot) = f_N(i,\cdot)$ is continuous for all i, and the continuity of v_n can be proved by induction as follows. Assume v_{n+1} to be continuous. From (3.10) and (3.17) we derive

$$v_n(i,x_0) = f_n(i,x_0) + c_n(i,\hat{u}(i,x_0)) + F_{n+1}(v_{n+1})(i,x_0+\hat{u}(i,x_0))$$

and, because for $x = x_1$ the infimum in (3.10) is not necessarily attained at $\hat{u}(i,x_0)$, we have

$$v_n(i,x_1) \le f_n(i,x_1) + c_n(i,\hat{u}(i,x_0)) + F_{n+1}(v_{n+1})(i,x_1+\hat{u}(i,x_0)).$$

Thus,

$$\begin{aligned} v_n(i,x_1) - v_n(i,x_0) &\le f_n(i,x_1) - f_n(i,x_0) \\ &+ F_{n+1}(v_{n+1})(i,x_1+\hat{u}(i,x_0)) - F_{n+1}(v_{n+1})(i,x_0+\hat{u}(i,x_0)), \end{aligned}$$

which, in view of the continuity of f_n and $v_{n+1} \in \mathbb{L}^\gamma$ and (3.1), yields

$$\limsup_{x_1\to x_0} \ v_n(i,x_1) - v_n(i,x_0) \le 0.$$

On the other hand, we have already proved that v_n is l.s.c., which means

$$\liminf_{x_1 \to x_0} v_n(i, x_1) - v_n(i, x_0) \geq 0.$$

Therefore, v_n is continuous. □

To solve the problem of minimizing $J_0(i, x; U)$, let us define

$$\begin{cases} \hat{x}_0 = x, \\ \hat{u}_n = \hat{u}_n(i_n, \hat{x}_n),\ n = 0, \ldots, N-1, \\ \hat{x}_n + 1 = \hat{x}_n + \hat{u}_n - \xi_n,\ n = 0, \ldots, N-1, \\ i_n,\ n = 0, \ldots, N,\ \text{Markov chain with transition matrix } P, \\ i_0 = i, \end{cases}$$

where $\hat{u}_n(i, x)$ is a Borel function for which the infimum in (3.10) is attained for any i and x.

THEOREM 3.2 (Verification Theorem) *The policy* $\hat{U} = (\hat{u}_0, \hat{u}_1, \ldots, \hat{u}_{N-1})$ *minimizes* $J_0(i, x; U)$ *over the class* $\mathcal{U}$ *of all admissible decisions. Moreover,*

$$v_0(i, x) = \min_{U \in \mathcal{U}} J_0(i, x; U). \tag{3.18}$$

Proof. Let $U = (u_0, \ldots, u_{N-1})$ be any admissible decision with the corresponding trajectory $(x_0, \ldots, x_{N-1})$. Without loss of generality, we may assume that $\mathsf{E}c_n(i_n, u_n) < \infty$, $\mathsf{E}f_n(i_n, x_n) < \infty$, $n \in \langle 0, N-1 \rangle$, and $\mathsf{E}f_N(i_N, x_N) < \infty$. Otherwise, $J_0(i, x; U) = \infty$ and U cannot be optimal since $J_0(i, x; \mathbf{0}) < \infty$ in view of (3.1) and Assumptions (i)-(iii).

Because $v_n(i_N, x_N) = f_N(i_N, x_N)$, it follows that $\mathsf{E}v_N(i_N, x_N) < \infty$. Using arguments analogous to those in the proof of Theorem 2.2, we proceed by induction. Assume that $\mathsf{E}v_{n+1}(i_{n+1}, x_{n+1}) < \infty$. Using property (vii) in Section B.2, (3.7), and (3.8), we obtain (see details leading to (2.15))

$$\mathsf{E}\{v_{n+1}(i_{n+1}, x_{n+1}) | \mathcal{F}^n\} = F_{n+1}(v_{n+1})(i_n, x_n + u_n) \text{ a.s.} \tag{3.19}$$

Since U is admissible but not necessarily optimal, we can use (3.10) to assert that

$$v_n(i_n, x_n) \leq f_n(i_n, x_n) + c_n(i_n, u_n) + \alpha F_{n+1}(v_{n+1})(i_n, x_n + u_n) \text{ a.s.}$$

Then from the relation (3.19), we can derive

$$v_n(i_n, x_n) \leq f_n(i_n, x_n) + c_n(i_n, u_n) + \alpha \mathsf{E}\{v_{n+1}(i_{n+1}, x_{n+1}) | \mathcal{F}^n\}.$$

Taking expectation of both sides of the above inequality, we obtain

$$\begin{aligned}\alpha^n \mathsf{E} v_n(i_n, x_n) &\leq \alpha^n \mathsf{E}(f_n(i_n, x_n) + c_n(i_n, x_n)) \\ &\quad + \alpha^{n+1} \mathsf{E}(v_{n+1}(i_{n+1}, x_{n+1})). \qquad (3.20)\end{aligned}$$

We can conclude recursively that $\mathsf{E} v_n(i_n, x_n) < \infty$ for all $n \in \langle 0, N \rangle$ and that (3.20) holds. Then, by summing (3.20) from 0 to $N-1$ and canceling identical terms on both sides, we obtain

$$v_0(i, x) \leq J_0(i, x; U). \qquad (3.21)$$

Consider now the decision $\hat{U}$. Using the definition of $\hat{u}_n(i_n, x)$ as the Borel function for which the infimum in (3.10) is attained, and proceeding as above, we can obtain

$$\begin{aligned}&\alpha^n \mathsf{E} v_n(i_n, \hat{x}_n) \\ &\quad = \alpha^n \mathsf{E}(f_n(i_n, \hat{x}_n) + c_n(i_n, \hat{u}_n)) + \alpha^{n+1} \mathsf{E}(v_{n+1}(i_{n+1}, \hat{x}_{n+1})).\end{aligned}$$

Note that $\hat{x}_0 = x$ is deterministic and $v_0(i, x) \in \mathbb{L}^\gamma$. Therefore, $\mathsf{E} v_0(i_0, \hat{x}_0) = v_0(i, x) < \infty$, and furthermore, it can be shown recursively that $\mathsf{E} c_n(i_n, \hat{u}_n) < \infty$, $n \in \langle 0, N{-}1 \rangle$ and $\mathsf{E} f_n(i_n, \hat{x}_n) < \infty$, $\mathsf{E} v_n(i_n, \hat{x}_n) < \infty$, $n \in \langle 0, N \rangle$. Adding up for n from 0 to $N{-}1$ and canceling terms, we get

$$v_0(i, x) = J_0(i, x; \hat{U}).$$

This and the inequality (3.21) complete the proof. □

3.4. Nonstationary Discounted Infinite Horizon Problem

In this section, we consider an infinite horizon version of the model formulated in Section 3.2. We require that the Assumptions (i) and (ii) hold with $N = \infty$, and that $i_0, i_1, \ldots$ is a Markov chain with the same transition matrix P. We set $N = \infty$, replace $\langle n, N \rangle$ by $\langle n, \infty \rangle$, replace the admissible decision in (3.4) by

$$U = (u_n, u_{n+1}, \ldots), \qquad (3.22)$$

and replace (3.5) by the objective function

$$J_n(i, x; U) = \sum_{k=n}^{\infty} \alpha^{k-n} \mathsf{E}[c_k(i_k, u_k) + f_k(i_k, x_k)], \qquad (3.23)$$

where α is the given discount factor, $0 < \alpha < 1$. The dynamic programming equations for each $i \in \mathbb{I}$ and $n = 0, 1, 2, \ldots$ can be written as

$$v_n(i,x) = f_n(i,x) + \inf_{u \geq 0} \{c_n(i,u) + \alpha F_{n+1}(v_{n+1})(i, x+u)\}. \tag{3.24}$$

In what follows, we will show that there exists an $\mathbb{L}^\gamma$-solution of (3.24), which is the value function of the infinite horizon problem. Moreover, the decision, for which the infimum in (3.24) is attained, is an optimal feedback policy.

First, let us examine the finite horizon approximation $J_{n,m}(i,x;U)$, $m \geq 1$, of (3.23). The approximation is obtained by the first m-period truncation of the infinite horizon problem of minimizing $J_n(i,x;U)$, i.e.,

$$J_{n,m}(i,x;U) = \sum_{k=n}^{n+m-1} \mathsf{E}[c_l(i_k, u_k) + f_k(i_k, x_k)]\alpha^{k-n}. \tag{3.25}$$

Let $v_{n,m}(i,x)$ be the value function of the truncated problem with no penalty cost in the last period, i.e.,

$$v_{n,m}(i,x) = \inf_U J_{n,m}(i,x;U). \tag{3.26}$$

Since the truncated problem is a finite horizon problem defined on the interval $\langle n, n+m\rangle$, Theorems 3.1 and 3.2 apply. Therefore, its value function can be obtained by solving the corresponding dynamic programming equations

$$\begin{aligned} v_{n,m+1}(i,x) &= f_n(i,x) + \inf_{u \geq 0} \{c_n(i,u) \\ &\qquad + \alpha F_{n+1}(v_{n+1,m})(i, x+u)\}, \end{aligned} \tag{3.27}$$

$$v_{n+m,0}(i,x) = 0. \tag{3.28}$$

Moreover, $v_{n,0}(i,x) = 0$ and

$$v_{n,m}(i,x) = \min_U J_{n,m}(i,x;U).$$

Next we will show that an *a priori* upper bound on $\inf J_n(i,x;U)$ can be easily constructed. Let us define

$$w_n(i,x) = J_n(i,x;\mathbf{0}), \tag{3.29}$$

where $\mathbf{0} = \{0, 0, \ldots\}$ is the policy that never orders anything. Then, since no production costs are incurred in view of Assumption (i), we have

$$w_n(i,x) = f_n(i,x) + \mathsf{E}\Big[\sum_{k=n+1}^{\infty} \alpha^{k-n} f_k(i_k, x-(\xi_n+\xi_{n+1}+\cdots+\xi_{k-1})\Big| i_n = i\Big]. \quad (3.30)$$

LEMMA 3.1 *$w_n(i,x)$ is well defined and $w_n(i,x) \in \mathbb{L}^\gamma$.*

Proof. On account of $f_n(i,x) \in \mathbb{L}^\gamma$, it is sufficient to show that

$$\mathsf{E}\Big[\sum_{k=n+1}^{\infty} \alpha^{k-n} f_k(i_k, x-(\xi_n+\xi_{n+1}+\cdots+\xi_{k-1}))\Big| i_n = i\Big] \in \mathbb{L}^\gamma.$$

Assumption (ii) yields

$$\begin{aligned}
&\mathsf{E}\Big[\sum_{k=n+1}^{\infty} \alpha^{k-n} f_k(i_k, x-(\xi_n+\xi_{n+1}+\cdots+\xi_{k-1}))\Big| i_n = i\Big] \\
&\le \bar{f}\sum_{k=n+1}^{\infty} \alpha^{k-n}(1+\mathsf{E}\Big[|x-(\xi_n+\xi_{n+1}+\cdots+\xi_{k-1})|^\gamma\Big| i_n = i\Big]) \\
&= \frac{\bar{f}\alpha}{1-\alpha} + \bar{f}\sum_{k=n+1}^{\infty} \alpha^{k-n}\mathsf{E}\Big[|x-(\xi_n+\xi_{n+1}+\cdots+\xi_{k-1})|^\gamma\Big| i_n = i\Big].
\end{aligned}$$

Note that it follows from (3.1) that $\mathsf{E}(\xi_k^\gamma | i_k = i) \le D+1$ for all k and i. Let $\overline{M}_x = \max\{|x|^\gamma, D+1\}$. Now let us consider the argument of the sum for a fixed $k \ge n+1$. Let $\Pi_k := \{(i_n,\ldots,i_k) : i_n,\ldots,i_k \in \mathbb{I} \text{ and } i_n = i\}$ be the set of all combinations of demand states in periods n through k for a given initial state $i_n = i$. Note that for a given sequence of demand states, the one-period demands are independent. Then we have,

$$\begin{aligned}
&\mathsf{E}\Big[|x-(\xi_n+\cdots+\xi_{k-1})|^\gamma\Big| i_n = i\Big] \\
&= \sum_{\pi\in\Pi_k} \mathsf{E}\Big[|x-(\xi_n+\cdots+\xi_{k-1})|^\gamma\Big| (i_n,\ldots,i_k) = \pi\Big] \\
&\qquad \times \mathsf{P}((i_n,\ldots,i_k) = \pi) \\
&\le \sum_{\pi\in\Pi_k} (k-n+1)^\gamma \mathsf{E}\Big[|x|^\gamma + |\xi_n|^\gamma + \cdots + |\xi_{k-1})|^\gamma\Big| (i_n,\ldots,i_k) = \pi\Big] \\
&\qquad \times \mathsf{P}((i_n,\ldots,i_k) = \pi) \\
&\le \sum_{\pi\in\Pi_k} (k-n+1)^\gamma (n-k)\bar{M}_x \mathsf{P}((i_n,\ldots,i_k) = \pi) \\
&= (k-n+1)^\gamma (k-n)\bar{M}_x.
\end{aligned}$$

Therefore,

$$\mathsf{E}\Big[\sum_{k=n+1}^{\infty}\alpha^{k-n}f_k(i_k, x-(\xi_n+\cdots+\xi_{k-1}))\Big|i_n=i\Big]$$

$$\leq \bar{f}\Big(\frac{\alpha}{1-\alpha}+\bar{M}_x\sum_{k=n+1}^{\infty}\alpha^{k-n}(k-n+1)^{\gamma}(k-n)\Big)<\infty.$$

Consequently, $w_n(i,x)<\infty$. In view of Theorem A.1.8, $w_n(i,x)$ is also l.s.c. as the sum of nonnegative l.s.c. functions. Moreover, because $\bar{M}_x = |x|^{\gamma}$ for $|x|^{\gamma}\geq D+1$, $w_n(i,x)$ is at most of polynomial growth with power γ. Thus, we have $w_n(i,x)\in\mathbb{L}^{\gamma}$. □

We can now state the following result for the infinite horizon problem.

THEOREM 3.3 *Let Assumptions* (i)-(ii) *and* (3.1) *hold. Then, we have*

$$0 = v_{n,0}\leq v_{n,1}\leq\ldots\leq v_{n,m}\leq w_n \tag{3.31}$$

and

$$v_{n,m}\uparrow v_n\in\mathbb{B}^{\gamma}, \tag{3.32}$$

where v_n is a solution of (3.24) *in $\mathbb{L}^{\gamma}$. Furthermore, there exists $\hat{U}=\{\hat{u}_n,\hat{u}_{n+1},\ldots\}$ for which the infimum in* (3.24) *is attained, and $\hat{U}$ is an optimal feedback policy, i.e.,*

$$v_n(i,x)=\min_U J_n(i,x;U)=J_n(i,x;\hat{U}). \tag{3.33}$$

Proof. By definition, $v_{n,0}=0$. Let $\tilde{U}_{n,m}=\{\tilde{u}_n,\tilde{u}_{n+1},\ldots,\tilde{u}_{n+m-1}\}$ be a minimizer of (3.25). Thus,

$$\begin{aligned} w_n(i,x) &= J_n(i,x;\mathbf{0})\geq J_{n,m}(i,x;\mathbf{0})\\ &\geq v_{n,m}(i,x)=J_{n,m}(i,x;\tilde{U}_{n,m})\geq J_{n,m-1}(i,x;\tilde{U}_{n,m-1})\\ &\geq \min_U J_{n,m-1}(i,x;U)=v_{n,m-1}(i,x). \end{aligned}$$

This proves (3.31). Moreover, it follows from (3.31) that there is a function $v_n(i,x)$ such that

$$v_{n,m}(i,x)\uparrow v_n(i,x)\leq w_n(i,x). \tag{3.34}$$

Next, we will show that the functions v_n satisfy the dynamic programming equations (3.24). Observe from (3.27) and (3.31) that for each m, we have

$$v_{n,m}(i,x)\leq f_n(i,x)+\inf_{u\geq 0}\{c_n(i,u)+\alpha F_{n+1}(v_{n+1,m})(i,x+u)\}.$$

Thus, in view of (3.34), we can replace $v_{n+1,m}$ by v_n on the RHS of the above inequality and then pass to the limit on the LHS as $m \to \infty$ to obtain

$$v_n(i,x) \leq f_n(i,x) + \inf_{u \geq 0}\{c_n(i,u) + \alpha F_{n+1}(v_{n+1})(i,x+u)\}. \tag{3.35}$$

Let the infimum in (3.27) be attained at $\hat{u}_{n,m}$. In order to obtain the reverse inequality, we first prove that $\hat{u}_{n,m}(i,x)$ is uniformly bounded with respect to m.

In the proof of Theorem 3.1 we showed that for $|x| \leq M$, there is an $\bar{u}^M_{n,i}$ such that $\hat{u}_{n,N-n}(i,x) \leq \bar{u}^M_{n,i}$. Furthermore, if we replace v_{n+1} by w_{n+1} in (3.12) and follow the same line of arguments as in the proof of Theorem 3.1, we obtain an upper bound $\bar{\bar{u}}^M_{n,i}$, which does not depend on the horizon N. Therefore, we can conclude that

$$\bar{u}_n(i,x) := \bar{\bar{u}}^M_{n,i} \text{ is an upper bound for } \hat{u}_{n,m}(i,x) \text{ if } M-1 < |x| \leq M, \tag{3.36}$$

independent of m.

For $l > m$, we see from (3.27)that

$$\begin{aligned} v_{n,l+1}(i,x) &= f_n(i,x) + c_n(i,\hat{u}_{n,l}(x)) \\ &\qquad + \alpha F_{n+1}(v_{n+1,l})(i,x+\hat{u}_{n,l}(x)) \\ &\geq f_n(i,x) + c_n(i,\hat{u}_{n,l}(x)) \\ &\qquad + \alpha F_{n+1}(v_{n+1,m})(i,x+\hat{u}_{n,l}(x)). \end{aligned} \tag{3.37}$$

Fix m and let $l \to \infty$. In view of (3.36), we can choose a sequence of periods l' such that

$$\hat{u}_{n,l'}(i,x) \to \tilde{u}_n(i,x) \text{ for } l' \to \infty. \tag{3.38}$$

We then conclude from (3.37) and Fatou's Lemma (Lemma B.1.1) that

$$\begin{aligned} v_n(i,x) &\geq f_n(i,x) + \lim_{l' \to \infty} c_n(i,\hat{u}_{n,l'}(i,x)) \\ &\qquad + \alpha \liminf_{l' \to \infty} F_{n+1}(v_{n+1,m})(i,x+\hat{u}_{n,l'}(i,x)), \\ &\geq f_n(i,x) + \lim_{l' \to \infty} c_n(i,\hat{u}_{n,l'}(i,x)) \\ &\qquad + \alpha \sum_{j=1}^{L} p_{ij} \int_0^\infty (\liminf_{l' \to \infty} v_{n+1,m}(i,x+\hat{u}_{n,l'}(i,x)-\xi)) d\Phi_{i,n}(\xi). \end{aligned}$$

Since $v_{n+1,m}$ and c_n are l.s.c., we can, in view of (3.38), pass to the limit in the argument of these functions to obtain

$$\begin{aligned} v_n(i,x) &\geq f_n(i,x) + c_n(i,\tilde{u}_n(i,x)) + \alpha F_{n+1}(v_{n+1,m})(i,x+\tilde{u}_n(i,x)), \\ &\geq f_n(i,x) + \inf_{u \geq 0}\{c_n(i,u) + \alpha F_{n+1}(v_{n+1,m})(i,x+u)\}. \end{aligned} \tag{3.39}$$

This, along with (3.35) and (3.34), proves (3.32).

From Theorem A.1.8, it follows that v_n is l.s.c., as the monotone limit of l.s.c. functions $v_{n,m}$ as $m \to \infty$. Also, since v_n is bounded by a function w_n of polynomial growth, we have $v_n \in \mathbb{L}^\gamma$.

Because $w_n \geq v_n \geq f_n$, it is clear that $\bar{u}_n(i,x)$ defined in (3.36) is also an upper bound for the minimizer in (3.24). Therefore, there exists a Borel map $\hat{u}_n(i,x)$ such that

$$\begin{aligned} & c_n(i, \hat{u}_n(i,x)) + \alpha F_{n+1}(v_{n+1})(i, x + \hat{u}_n(i,x)) \\ & \quad = \inf_{u \geq 0} \{c_n(i,u) + \alpha F_{n+1}(v_{n+1})(i, x+u)\}. \end{aligned} \tag{3.40}$$

With that in mind, we can use (3.24) to obtain

$$\begin{aligned} & \mathsf{E}[v_k(i_k, x_k)] \\ & \quad = \mathsf{E}[f_k(i_k, x_k) + c_k(i_k, \hat{u}_k)] + \alpha \mathsf{E}[F_{k+1}(v_{k+1})(i_k, x_k + \hat{u}_k)] \\ & \quad = \mathsf{E}[f_k(i_k, x_k) + c_k(i_k, \hat{u}_k] + \alpha \mathsf{E}[v_{k+1}(i_{k+1}, x_{k+1})], k = 0, 1, 2, \ldots. \end{aligned}$$

Multiplying by α^{k-n}, summing from n to $N-1$, and canceling terms yield

$$v_n(i,x) \geq \mathsf{E}\Big[\sum_{k=n}^{N-1} \alpha^{k-n}(c_k(i_k, \hat{u}_k) + f_k(i_k, \hat{x}_k))\Big] + \alpha^{N-n} \mathsf{E}[v_N(i_N, \hat{x}_N)].$$

Letting $N \to \infty$, we conclude

$$v_n(i,x) \geq J_n(i,x;\hat{U}). \tag{3.41}$$

From Theorem 3.2 we know that $v_{n,k}(i,x) \leq J_{n,k}(i,x;U)$ for any policy U, and we let $k \to \infty$ to obtain

$$v_n(i,x) \leq J_n(i,x;U) \text{ for any admissible } U. \tag{3.42}$$

Together, inequalities (3.41) and (3.42) imply

$$v_n(i,x) = J_n(i,x;\hat{U}) = \min_U J_n(i,x;U),$$

which completes the proof. □

Before we prove the optimality of an (s, S)-type policy for the nonstationary finite and infinite horizon problems, we should note that Theorem 3.3 does not imply uniqueness of the solution to the dynamic programming equations (3.27) and (3.28). There may be other solutions. Moreover, one can show that the value function is the minimal positive solution of (3.27) and (3.28). It is also possible to obtain a uniqueness proof under additional assumptions. For our purpose, however, it is sufficient to have the results of Theorem 3.3.

3.5. Optimality of (s, S)-type Ordering Policies

The existence and optimality of a *feedback* (or Markov) policy $\hat{u}_n(i,x)$ was proved in Theorems 3.1 and 3.2. We now make additional assumptions to further characterize the optimal feedback policy.

Let us assume that for any demand state i,

$$f_n(i,x) \quad \text{is convex with respect to } x, \quad n = 0,1,\ldots,N, \tag{3.43}$$

$$c_n(i,u) = \begin{cases} 0, & u = 0, \\ K_n^i + c_n^i \cdot u, & u > 0, \end{cases} \tag{3.44}$$

where $c_n^i \geq 0$ and $K_n^i \geq 0$, $n = 0,1,\ldots,N-1$, and

$$K_n^i \geq \alpha \bar{K}_{n+1}^i \equiv \alpha \sum_{j=1}^{L} p_{ij} K_{n+1}^j, \quad n = 0,1,\ldots,N. \tag{3.45}$$

It should be noted that (3.43) implies that $f_n(i,\cdot)$, for any i and $n = 0,1,\ldots,N$, is continuous on $\mathbb{R}$.

Remark 3.2 Assumptions (3.43)–(3.45) reflect the usual structure of costs to prove optimality of an (s,S)-type policy.

Theorem 3.4 *Let N be finite. Let Assumptions* (i)-(iii), (3.1), (3.9), *and* (3.43)-(3.45) *hold. Then, there exists a sequence of numbers $s_{n,i}$, $S_{n,i}$, $n = 0,\ldots,N-1$, $i = 1,\ldots,L$, with $s_{n,i} \leq S_{n,i}$, such that the optimal feedback policy is*

$$\hat{u}_n(i,x) = \begin{cases} S_{n,i} - x, & x \leq s_{n,i}, \\ 0, & x > s_{n,i}. \end{cases} \tag{3.46}$$

Proof. The dynamic programming equations (3.10) and (3.11) can be written as

$$\begin{aligned} v_n(i,x) &= f_n(i,x) - c_n^i x + h_n(i,x), \quad 0 \leq n \leq N-1, \ i = 1,\ldots,L, \\ v_N(i,x) &= f_N(i,x), \qquad\qquad\qquad\qquad\quad i = 1,\ldots,L, \end{aligned}$$

where

$$h_n(i,x) = \inf_{y \geq x} [K_n^i \mathbb{I}_{y>x} + z_n(i,y)], \tag{3.47}$$

$$z_n(i,y) = c_n^i y + \alpha F_{n+1}(v_{n+1})(i,y). \tag{3.48}$$

From (3.10), we have $v_n(i,x) \geq f_n(i,x)$. This inequality, along with (3.9), ensures for $n = 1,2,\ldots,N-1$ and $i = 1,2,\ldots,L$ that

$$z_n(i,x) \to +\infty \text{ as } x \to \infty. \tag{3.49}$$

Furthermore, from the last part of Theorem 3.1, it follows that v_{n+1} is continuous; therefore $z_n(i,x)$ is continuous.

In order to obtain (3.46), we need to prove that $z_n(i,x)$ is K_n^i-convex. This is done by induction. First, $v_N(i,x)$ is convex by definition and therefore, K-convex for any $K \geq 0$. Let us now assume that for a given $n \leq N-1$ and i, $v_{n+1}(i,x)$ is K_{n+1}^i-convex. By Assumption (3.45), it is easy to see that $z_n(i,x)$ is $\alpha \bar{K}_{n+1}^i$-convex, hence also K_n^i-convex. Then, Theorem C.2.3 implies that $h_n(i,x)$ is K_n^i-convex. Therefore, $v_n(i,x)$ is K_n^i-convex. This completes the induction argument.

Thus, it follows that $z_n(i,x)$ is K_n^i-convex for each n and i. In view of (3.49), we apply Theorem C.2.3 to obtain the desired $s_{n,i}$ and $S_{n,i}$. According to Theorem 3.2 and the continuity of z_n, the optimal feedback policy defined in (3.46) is optimal. □

THEOREM 3.5 *Let Assumptions* (i)-(ii), (3.1), (3.9), *and* (3.43)–(3.45) *hold for the cost functions for the infinite horizon problem. Then, there exists a sequence of numbers $s_{n,i}, S_{n,i}$, $n = 0, 1, \ldots$, with $s_{n,i} \leq S_{n,i}$ for each $i \in \mathbb{I}$, such that the feedback policy*

$$\hat{u}_n(i,x) = \begin{cases} S_{n,i} - x, & x < s_{n,i}, \\ 0, & x \geq s_{n,i}, \end{cases} \tag{3.50}$$

is optimal.

Proof. Let v_n denote the value function. Define the functions z_n and h_n as above. We know that $z_n(i,x) \to \infty$ as $x \to +\infty$ and $z_n(i,x) \in \mathbb{L}^\gamma$ for all n and $i = 1, 2, \ldots, L$.

We now prove that v_n is K_n-convex. Using the same induction as in the proof of Theorem 3.4, we can show that $v_{n,m}(i,x)$, defined in (3.26), is K_n^i-convex. This induction is possible since we know that $v_{n,m}(i,x)$ satisfies the dynamic programming equations (3.27) and (3.28). It is clear from the definition of K-convexity that this property is preserved under monotone limit procedures. Thus, the value function $v_n(i,x)$, which is the limit of $v_{n,m}(i,x)$ as $m \to \infty$, is K_n^i-convex.

From Theorem 3.3 we know that v_n satisfies the dynamic programming equations (3.27) and (3.28). Therefore, we can obtain an optimal feedback policy $\hat{U} = \{\hat{u}_n, \hat{u}_{n+1}, \ldots\}$, for which the infimum in (3.27) is attained. Because z_n is K_n^i-convex and l.s.c., $\hat{u}_n$ can be expressed as in (3.50). □

REMARK 3.3 It is important to emphasize the difference between the (s,S) policies defined in (3.46) and (3.50). In (3.46), an order is placed when the inventory level is s or below, whereas in (3.50) an order is placed only when the inventory is strictly below s. Most of the literature uses the policy type (3.46). While (3.46) in Theorem 3.4 can be replaced by (3.50) on account of the continuity of z_n, it is not possible to replace (3.50) in Theorem 3.5 by (3.46), since z_n is proved only to be l.s.c.

REMARK 3.4 In the stationary infinite horizon discounted cost case discussed in the next section, we are able to prove that the value function is locally Lipschitz, and therefore continuous. The proof is provided in Chapter 5, Lemma 5.3. Thus, in this case, policies of both types (3.50) and (3.46) are optimal.

3.6. Stationary Infinite Horizon Problem

If the cost functions, as well as the distributions of the demands, do not explicitly depend on time, i.e., for each k

$$c_k(i,u) = c(i,u), \quad f_k(i,x) = f(i,x), \quad \text{and} \quad \Phi_{i,k} = \Phi_i,$$

then it can be easily shown that the value function $v_n(i,x)$ does not depend on n. In what follows, we will denote the value function of the stationary discounted cost problem by $v^\alpha(\cdot,\cdot)$, in order to emphasize the dependence on the discount factor α. In the same manner as in Section 3.4, it can be proved that the function v^α satisfies the dynamic programming equation

$$v^\alpha(i,x) = f(i,x) + \inf_{u \geq 0}\{c(i,u) + \alpha F(v^\alpha)(i,x+u)\}, \tag{3.51}$$

where F is the same as F_{n+1}, defined in (3.8), i.e.,

$$Fb(i,y) = \sum_{j=1}^{L} p_{ij} \int_0^\infty b(j,y-\xi)d\Phi_i(\xi),$$

for $b \in \mathbb{B}^\gamma$.

Furthermore, for any α, $0 < \alpha < 1$, there is a *stationary optimal feedback policy* $U^\alpha = (u^\alpha(i,x), u^\alpha(i,x), \ldots)$, where $u^\alpha(i,x)$ is the minimizer on the RHS of (3.51). Moreover, if the cost functions also satisfy the Assumptions (3.43)–(3.45) introduced in Section 3.5, then we can obtain pairs (s_i^α, S_i^α) such that either of the (s_i^α, S_i^α)-policies of types (3.50) and (3.46) is optimal; (see Remark 3.4).

3.7. Concluding Remarks and Notes

This chapter, based on Beyer and Sethi (1997) and Beyer *et al.* (1998), generates the discounted infinite horizon inventory model involving fixed costs that have appeared in the literature, to allow for unbounded demand and costs with polynomial growth. We have shown the existence of an optimal Markov policy, and that this can be a state-dependent (s, S) policy.

This chapter makes several specific contributions. It extends the proofs of existence and verification of optimality in the discounted cost case given in Chapter 2, to allow for more general costs including l.s.c. surplus cost with polynomial growth.

Some problems of theoretical interest remain open. One might want to show that the value function in the discounted nonstationary infinite horizon case is continuous if the surplus cost function is continuous.

Chapter 4

DISCOUNTED COST MODELS WITH LOST SALES

4.1. Introduction

In the literature of stochastic inventory models, there are two different assumptions about the excess demand unfilled from existing inventories: the backlog assumption and the lost sales assumption. The former is more popular in the literature, partly because historically the inventory studies started with spare parts inventory management problems in military applications, where the backlog assumption is realistic. However, in many other business situations, it is quite often that demand that cannot be satisfied on time is lost. This is particularly true in a competitive business environment. For example, in many retail establishments such as a supermarket or a department store, a customer chooses a competitive brand and goes to another store if his/her preferred brand is out of stock.

In the presence of fixed ordering costs in inventory models under either assumption, an important issue has been to establish the optimality of (s, S)-type policies. However, in comparison to many classical and recent papers dealing with this issue in the backlog case, there are only a few that treat the lost sales case. These include Veinott (1966), Shreve (1976), and Bensoussan *et al.* (1983). This is perhaps because the proofs of the results in the lost sales case are usually more complicated than those in the backlog case. Shreve (1976) and Bensoussan *et al.* (1983) establish the optimality of an (s, S)-type policy by using the concept of K-convexity. Veinott (1966) provides a different proof for the optimality of (s, S)-type policies in the lost sales case. His proof is based on a different set of assumptions which neither implies nor is implied by those

D. Beyer et al., *Markovian Demand Inventory Models*, International Series in Operations Research & Management Science 108,
DOI 10.1007/978-0-387-71604-6_4,

used in Shreve (1976) and Bensoussan *et al.* (1983). It should be noted that all of these results are obtained under the condition of zero leadtime.

Many efforts have been made to incorporate various realistic features in inventory models. However, most of them are carried out under the backlog assumption. One such feature is that of Markovian demand, which is the subject of this book. In the previous two chapters, as in many of the references cited therein, the backlog assumption is used in all the analyses. As the lost sales situation is often the case in competitive markets, it is interesting and worthwhile to extend the Markovian demand model to allow for this situation.

This chapter studies a Markovian demand inventory model used when any unsatisfied demand in a period is lost, whereas in Chapters 2 and 3, such unsatisfied demands are backlogged. The plan of this chapter is as follows. In Section 4.2, we formulate our basic lost sales model with Markovian demands, provide the dynamic programming equations, and state the existence results. In Section 4.3, we show that state-dependent (s, S) policies are optimal for the lost sales case of the Markovian demand model using a K-convexity result associated with the cost functions in the lost sales case. The proof of the K-convexity result is provided in Section C.2. Furthermore, our Markovian demand model is formulated under a set of less restrictive assumptions than those in the independent demand models of Shreve (1976) and Bensoussan *et al.* (1983). In Section 4.4, we discuss extensions that incorporate various realistic features and constraints, such as supply uncertainty, service levels, and storage capacities. An infinite horizon stationary model is also presented. Numerical results are presented in Section 4.5. We conclude the chapter with some remarks in Section 4.6.

4.2. Formulation of the Model

Consider an inventory problem over a finite horizon. The demand in each period is assumed to be a random variable defined on a given probability space, and not necessarily identically distributed. To precisely define the demand process, we consider a finite collection of demand states labeled $i \in \mathbb{I} = \{1, 2, \ldots, L\}$, and let i_k denote the demand state observed at the beginning of period k. We assume that i_k, $k \in \langle 0, N\rangle$, is a Markov chain over $\mathbb{I}$, with the transition matrix $P = \{p_{ij}\}$. Thus,

$$0 \le p_{ij} \le 1, \;\; i \in \mathbb{I}, j \in \mathbb{I}, \;\; \text{and} \;\; \sum_{j=1}^{L} p_{ij} = 1, \;\; i \in \mathbb{I}.$$

Let the nonnegative random variable ξ_k denote the demand at the end of a given period $k \in \langle 0, N-1\rangle$. Demand ξ_k depends only on period

k and the demand state in that period, by which we mean that it is independent of past demand states and past demands. We denote its probability density by $\varphi_{i,k}(x)$, when the demand state $i_k = i$. We assume that

$$\mathsf{E}(\xi_k | i_k = i) = \int_0^\infty z\varphi_{i,k}(z)\, dz \le D < \infty, \; k \in \langle 0, N-1 \rangle, \; i \in \mathbb{I}. \quad (4.1)$$

At the beginning of period k, an order $u_k \ge 0$ is placed with the knowledge that the demand state is i_k and that the quantity u_k will be delivered at the end of period k, but before the period k demand ξ_k materializes. If the on-hand inventory x_k at the beginning of period k, plus the amount u_k delivered in period k, exceeds the demand, i.e., if $x_k + u_k \ge \xi_k$, then the demand is met and the remaining inventory is carried over to the next period as $x_{k+1} = x_k + u_k - \xi_k$. If $x_k + u_k < \xi_k$, then the part $\xi_k - x_k - u_k$ of the demand cannot be satisfied immediately and is assumed to be completely lost. In this case, the next period will be started with zero on-hand inventory. With the notation $a^+ = \max\{a, 0\}$ and $0 \le n \le N$, the model dynamics can be expressed as

$$\begin{cases} x_{k+1} = (x_k + u_k - \xi_k)^+, \; k \in \langle n, N-1 \rangle \\ x_n = x \ge 0, \\ i_k, \; k \in \langle n, N \rangle, \text{ follows a Markov chain with transition matrix } P, \\ i_n = i. \end{cases} \quad (4.2)$$

This describes the dynamics from period n onwards, given the inventory level x and demand state i in period n.

Let $\mathbb{B}_0$ denote the class of all continuous functions from $\mathbb{I} \times \mathbb{R}$ into $\mathbb{R}^+$, and the pointwise limits of sequences of these functions (see Feller (1971)), where $\mathbb{R} = (-\infty, \infty)$ and $\mathbb{R}^+ = [0, \infty)$. Note that it includes l.s.c. functions that arise in Section 4.4. Let $\mathbb{B}_1$ be the subspace of functions in $\mathbb{B}_0$ that are l.s.c. and of linear growth, i.e., $0 \le b(i, x) \le C_b(1 + |x|)$ for some $C_b > 0$.

Next, we define various costs that are involved and the assumptions they satisfy.

(i) For period $k = 0, \ldots, N-1$ and demand state $i = 1, \ldots, L$, let

$$\begin{aligned} K_k^i &= \text{the fixed order cost in period } k \text{ if } i_k = i, \; K_k^i \ge 0, \\ c_k^i &= \text{the variable order cost in period } k \text{ if } i_k = i, \; c_k^i \ge 0, \\ c_k(i, u) &= K_k^i \mathbb{1}_{u>0} + c_k^i u, \text{ the cost of ordering } u \text{ units in} \\ &\quad\; \text{period } k \text{ if } i_k = i. \end{aligned} \quad (4.3)$$

(ii) The surplus cost in period k when $i_k = i$ and $x_k = x$ is denoted by $f_k(i,x)$, and

$$\begin{cases} f_k(i,x) \in \mathbb{B}_1, \text{ is convex and nondecreasing in } x, \text{ and} \\ f_k(i,x) = 0, \ \forall x \le 0. \end{cases} \tag{4.4}$$

(iii) The shortage cost in period k when $i_k = i$ and $x_k = x$ is denoted by $q_k(i,x)$, and we have

$$\begin{cases} q_k(i,x) \in \mathbb{B}_1, \text{ is convex and nonincreasing in } x, \text{ and} \\ q_k(i,x) = 0, \ \forall x \ge 0. \end{cases} \tag{4.5}$$

(iv) Furthermore, for $k \in \langle 0, N-1 \rangle$,

$$c_k^i x + \mathsf{E} f_{k+1}(i, x - \xi_k) \to +\infty \quad as \quad x \to +\infty. \tag{4.6}$$

With costs thus defined, we can for any n, $0 \le n \le N$, write the objective function as

$$\begin{aligned} J_n(i,x;U) \quad = \quad & \mathsf{E}\Big[\sum_{k=n}^{N-1} \{c_k(i_k,u_k) + f_k(i_k,x_k) + q_k(i_k, x_k + u_k - \xi_k)\} \\ & + f_N(i_N, x_N) | i_n = i, x_n = x\Big], \end{aligned} \tag{4.7}$$

where $U = (u_n, \ldots, u_{N-1})$ is an admissible *history-dependent* or *nonanticipative* (order quantities) for the problem, and u_k is a function of i_k and x_k, $k \in \langle n, N-1 \rangle = \{n, \ldots, N-1\}$. The objective function $J_n(i,x;U)$ represents the cost to go from period n on with the initial conditions at period n being i and x under the decision U. With $\mathcal{U}$ denoting the class of all admissible decisions, we can define the value function $v_n(i,x)$ for the problem over $\langle n, N \rangle$ as

$$v_n(i,x) = \inf_{U \in \mathcal{U}} J_n(i,x;U). \tag{4.8}$$

The dynamic programming equations for the model are defined for $n \in \langle 0, N-1 \rangle$ and $i = 1, 2, \ldots, L$ as

$$\begin{aligned} v_n(i,x) \quad = \quad & f_n(i,x) + \inf_{u \ge 0} \{c_n(i,u) + \mathsf{E}[q_n(i, x + u - \xi_n) \\ & + v_{n+1}(i_{n+1}, (x + u - \xi_n)^+) | i_n = i]\}, \end{aligned} \tag{4.9}$$

and

$$v_N(i,x) = f_N(i,x), \qquad i = 1,2,\ldots,L. \tag{4.10}$$

The existence of an optimal feedback policy can be established by argument similar to those used in the backlog case treated in Chapter 2. Therefore, we will state them without proof.

THEOREM 4.1 (Verification Theorem) *The dynamic programming equations* (4.10) *and* (4.10) *define a sequence of functions* $v_n(i,x)$ *in* $\mathbb{B}_1$. *Furthermore, there exists a function* $\hat{u}_n(i,x)$ *in* $\mathbb{B}_1$, *which provides the infimum in* (4.10) *for any* x, *and* $\hat{U} = (\hat{u}_0, \hat{u}_1, \ldots, \hat{u}_{N-1})$ *is an optimal decision for the problem* $J_0(i,x;U)$. *Moreover,*

$$v_0(i,x) = \min_{U \in \mathcal{U}} J_0(i,x;U). \tag{4.11}$$

In simpler words, the theorem, called a verification theorem, means that there exists a policy in the class of all admissible (or history-dependent) policies, whose objective function value equals the value function defined by (4.8). In addition, there is a Markov (or feedback) policy which gives the same objective function value.

4.3. Optimality of (s,S)-type Ordering Policies

The optimality of (s,S)-type policies has been established for stochastic inventory models with various conditions on demand and cost functions. A key concept used in proving the optimality of (s,S)-type policies for standard models in the literature is K-convexity of a function, which was first utilized by Scarf (1960). Some useful properties of K-convex functions are given in Section C.2.

The traditional definition of K-convex functions defined on the real line is extended in Section C.2 to include functions defined on convex subsets of the real line. Existing results on the properties of K-convex functions are also generalized to allow for less restrictive assumptions and more realistic features arising in inventory models. Based on these results, the optimality of state-dependent (s,S) policies is established in Markovian demand inventory models for the full backlog case.

However, in the lost sales case, the truncation due to lost sales requires new conditions to be introduced to preserve K-convexity in the induction step of the proof to establish the optimality of (s,S)-type policies. The required K-convexity result is proved as Theorem C.2.4.

We now prove the optimality of (s,S)-type policies for the lost sales case with the help of Theorem C.2.4.

THEOREM 4.2 *In addition to Assumptions* (i)-(iv) *in Section* 4.2, *let for each* $i \in \mathbb{I}$ *and* $j \in \mathbb{I}$,

$$K_n^i \geq \bar{K}_{n+1}^i \equiv \sum_{j=1}^{L} p_{ij} K_{n+1}^j \geq 0, \; n \in \langle 0, N-1 \rangle, \tag{4.12}$$

and

$$q_n^{'-}(i,0) \leq f_{n+1}^{'+}(j,0) - c_{n+1}^j, \; n \in \langle 0, N-2 \rangle. \tag{4.13}$$

Then, there exists a sequence $s_{n,i}, S_{n,i}$, *with* $0 \leq s_{n,i} \leq S_{n,i}$, $j \in \mathbb{I}$ *and* $n \in \langle 0, N-1 \rangle$, *such that the feedback policy*

$$\hat{u}_n(i,x) = \begin{cases} S_{n,i} - x, & \text{for } x < s_{n,i}, \\ 0, & \text{for } x \geq s_{n,i}, \end{cases} \tag{4.14}$$

is optimal.

Proof. We can rewrite (4.10) as

$$\begin{aligned} v_n(i,x) &= f_n(i,x) - c_n^i x + \inf_{y \geq x} \{K_n^i \mathbb{1}_{y>x} + c_n^i y + \mathsf{E}[q_n(i, y-\xi_n) \\ &\quad + v_{n+1}(i_{n+1}, (y-\xi_n)^+) | i_n = i]\}, \quad x \geq 0, \\ &= f_n(i,x) - c_n^i x + H_n(i,x), \quad x \geq 0, \end{aligned} \tag{4.15}$$

where for $x \geq 0$,

$$H_n(i,x) = \inf_{y \geq x \geq 0} [K_n^i \mathbb{1}_{y>x} + Z_n(i,y)], \tag{4.16}$$

$$\begin{aligned} Z_n(i,x) &= c_n^i x + \mathsf{E}\left[q_n(i, x-\xi_n)\right. \\ &\quad \left. + v_{n+1}(i_{n+1}, (x-\xi_n)^+) | i_n = i\right]. \end{aligned} \tag{4.17}$$

Note that $v_n(i,x)$ is defined for $x \in \mathbb{R}^+ = [0, \infty)$. In view of Theorem C.2.3 with $A = 0$ and $B = +\infty$, we need only to prove that

$$\begin{aligned} &Z_n(i,x) \text{ is l.s.c. and } K_n^i\text{-convex on } \mathbb{R}^+ \text{ and} \\ &\qquad Z_n(i,x) \to +\infty \text{ as } x \to +\infty. \end{aligned} \tag{4.18}$$

This is because with (4.16), (4.18), Theorem C.2.2(ii), and Theorem C.2.3(iii), we can conclude $v_n(i,x)$ to be K_n^i-convex on $\mathbb{R}^+$. Since $v_N(i,x)$ in (4.10) is convex by definition of $f_N(i,x)$, $Z_{N-1}(i,x)$ is convex and the proof of the theorem follows from an induction argument beginning with the premise that $Z_{n+1}(i,x)$ is K_{n+1}^i-convex on $\mathbb{R}^+$.

With regards to (4.18), note that $Z_n \in \mathbb{B}_1$ since $v_n \in \mathbb{B}_1$. Furthermore, from Assumption (4.6) and the facts that $v_{n+1} \geq f_{n+1} \geq 0$ in (4.17) and $q_n(i,x)$ is nonincreasing in x, we have

$$\begin{aligned} Z_n(i,x) \;\geq\; & c_n^i x + q_n(i,0) + \mathsf{E}\left[f_{n+1}(i_{n+1}, (x-\xi_n)^+) | i_n = i\right] \to +\infty, \\ & \text{as } x \to +\infty. \end{aligned}$$

Finally, to prove the K_n^i-convexity of Z_n on $\mathbb{R}^+$, let us define

$$Q_n(i,x) = q_n(i,x) + \bar{v}_{n+1}(i,x^+), \tag{4.19}$$

where

$$\bar{v}_{n+1}(i,x) = \mathsf{E}\left[v_{n+1}(i_{n+1},x)|i_n = i\right] = \sum_{j=1}^{L} p_{ij} v_{n+1}(j,x).$$

Thus, we have

$$\begin{aligned} Z_n(i,x) &= c_n^i x + \mathsf{E}[Q_n(i,x-\xi_n)|i_n = i] \\ &= c_n^i x + \int_0^\infty Q_n(i,x-\xi)\varphi_{i,n+1}(\xi)d\xi. \end{aligned}$$

Now we prove the K_n^i-convexity of Z_n on $\mathbb{R}^+$ by induction. It is sufficient to show that $Q_n(i,x)$ is $\bar{K}_{n+1}^i$-convex on $\mathbb{R}$ given the induction premise that $Z_{n+1}(i,x)$ is K_{n+1}^i-convex on $\mathbb{R}^+$ for each $j \in \mathbb{I}$.

We note that if $Z_{n+1}(i,x)$ is K_{n+1}^i-convex on $\mathbb{R}^+$, then $\bar{v}_{n+1}(i,x)$ is $\bar{K}_{n+1}^i$-convex for $x \in \mathbb{R}^+$ on account of Theorem C.2.3 and (4.3). In the following, we will show that $Q_n(i,x)$ is $\bar{K}_{n+1}^i$-convex on $\mathbb{R}$. This is achieved by showing that $q_n(i,x) + v_{n+1}(j,x^+)$ is K_{n+1}^j-convex on $\mathbb{R}$ for each j.

From the K_{n+1}^j-convexity of $Z_{n+1}(j,x)$ on $\mathbb{R}^+$ and Theorem C.2.3, there exist s_{n+1}^j and S_{n+1}^j, $S_{n+1}^j \geq s_{n+1}^j \geq 0$, such that

$$v_{n+1}(j,x) = f_{n+1}(j,x) - c_{n+1}^j x + \begin{cases} K_{n+1}^j + Z_{n+1}(j,S_{n+1}^j), & \text{if } x < s_{n+1}^j, \\ Z_{n+1}(j,x), & \text{if } x \geq s_{n+1}^j. \end{cases}$$

When $s_{n+1}^j > 0$, we in view of (4.13) have

$$v_{n+1}^{'+}(i,0) = f_{n+1}^{'+}(j,0) - c_{n+1}^j \geq q_n^{'-}(i,0).$$

According to Theorem C.2.4, $q_n(i,x) + v_{n+1}(j,x^+)$ is K_{n+1}^j-convex on $\mathbb{R}^+$.

When $s_{n+1}^j = 0$, we know that $v_{n+1}(j,x^+)$ is K_{n+1}^j-convex for $x \in \mathbb{R}^+$ given that $Z_{n+1}(j,x)$ is K_{n+1}^j-convex. Therefore, $Q_n(i,x) = q_n(i,x) + \bar{v}_{n+1}(i,x^+)$ is $\bar{K}_{n+1}^i$-convex on $\mathbb{R}^+$. This completes the proof for the theorem. □

REMARK 4.1 Assumption (4.6) is the same as (2.11) in Chapter 2; (see Remark 2.1 for its meaning).

REMARK 4.2 Assumption (4.13) means that the marginal shortage cost in one period is larger than or equal to the expected unit ordering cost less the expected marginal inventory holding cost in any state of the next period. If this condition does not hold, that is, if $-q_n^{\prime -}(i, 0) < \vec{c}_{n+1}^{\,i} - \vec{f}_{n+1}^{\prime +}(i, 0)$ for some i, a speculative retailer may find it attractive to meet a smaller part of the demand in period n than is possible from the available stock, carry the leftover inventories to period $n + 1$, and order a little less as a result in period $n + 1$ with the expectation that he will be better off. Thus, Assumption (4.13) rules out this kind of speculation on the part of the retailer. But such a speculative behavior is not allowed in our formulation of the dynamics (4.2) in any case, since the demand in any period must be satisfied to the extent of the availability of inventories. This suggests that it might be possible to prove Theorem 4.2 without (4.13). Moreover, our proof relies on the K-convexity of the value function, whereas this property is only a sufficient condition and not necessary for the optimality of an (s, S) policy.

Theorem 4.2 extends the analysis carried out in Chapter 2 to the lost sales case with zero leadtime and additional minor restrictions on the cost structure. In the backlog case, the mathematical analysis based on the zero leadtime assumption can be extended easily to the nonzero leadtime case by replacing the inventory level with the inventory position, which is the sum of the inventory on-hand and amounts on order. But the relationship between the inventory position, the inventory on-hand, and amounts on order is not straightforward in the lost sales case. Therefore, we cannot claim that the same results will hold for the lost sales case with nonzero leadtime.

4.4. Extensions

The model formulated and analyzed in Sections 4.2 and 4.3 can be extended to incorporate some additional constraints and realistic features that often arise in practice. It can be shown that (s, S)-type policies continue to remain optimal for the extended models. In the following, we briefly describe the new features in these extensions.

4.4.1 Supply Uncertainty

We model supply uncertainty by replacing the demand state i with a demand/supply state $i = (i^d, i^s)$, where i^d denotes the demand state and i^s denotes the supply state. We also redefine $\mathbb{I}$ to denote the set of all possible demand/supply states. Let r_k be a random variable representing the availability of the supply in period k. If the supply is available in period k, then $r_k = 1$; otherwise, $r_k = 0$. Therefore, the inventory balance equations are given by

$$x_{k+1} = (x_k + r_k \cdot u_k - \xi_k)^+, \; k \in \langle n, N-1 \rangle.$$

We further assume that the probability of supply being available in period k with the demand/supply state i is R_{ki}, i.e.,

$$\mathsf{P}(r_k = 1 | i_k = i) = R_{ki}, \quad k \in \langle 0, N \rangle, \; i \in \mathbb{I},$$

with $0 \le R_{ki} \le 1$. Accordingly, the dynamic programming equations of the value functions should be revised as

$$\begin{aligned} v_n(i,x) \;\; = \;\; & f_n(i,x) + (1 - R_{n,i})\mathsf{E}[q_n(i, x - \xi_{n+1}) \\ & + v_{n+1}(i_{n+1}, (x - \xi_n)^+) | i_n = i] \\ & + R_{n,i} \inf_{u \ge 0} \{c_n(i,u) + \mathsf{E}[q_n(i, x + u - \xi_n) \\ & + v_{n+1}(i_{n+1}, (x + u - \xi_n)^+) | i_n = i]\}. \end{aligned} \quad (4.20)$$

THEOREM 4.3 *In the inventory problem with the dynamic programming equations defined by* (4.20)*, the optimal policy is still of* (s,S)*-type.*

Proof. First, note that the first two items in the braces of (4.20) are independent of u. Also, in a way similar to the proof of Theorem 4.2, one can easily verify that the sum of these items is K_n^i-convex with respect to x. The rest of the proof is straightforward in view of the proof of Theorem 4.2. □

4.4.2 Storage and Service Level Constraints

Following the same definition in Section 2.8.2 of Chapter 2, we let $B < \infty$ be the storage capacity, and A_k^i be the safety stock level in period k with the demand state i. Thus, the inventory position is restricted by the lower bound A_k^i and the upper bound B such that

$$A_k^i \le x_k + u_k \le B. \quad (4.21)$$

The dynamic programming equations can be written as (2.22), where $z_n(i,y)$ is as in (2.22) and

$$h_n(i,x) = \inf_{y \ge x, A_n^i \le y \le B} [K_n^i \mathbb{1}_{y > x} + z_n(i,y)], \quad (4.22)$$

provided $A_n^i \le B$, $n \in \langle 0, N-1 \rangle, i \in \mathbb{I}$; if not, then there is no feasible solution, and $h_n(i,x) = \inf \emptyset \equiv \infty$. Furthermore, a verification theorem can also be obtained for the constrained case; (see Bertsekas and Shreve (1976)).

THEOREM 4.4 *Assume* $A_n^i \leq B$, $n \in \langle 0, N-1 \rangle, i \in \mathbb{I}$. *Then, there exists a sequence of numbers* $s_{n,i}, S_{n,i}$, $n \in \langle 0, N-1 \rangle, i \in \mathbb{I}$, *with* $s_{n,i} \leq S_{n,i}$ *and* $A_n^i \leq S_{n,i} \leq B$ *such that the optimal feedback policy is*

$$\hat{u}_n(i,x) = \begin{cases} S_{n,i} - x, & \text{for} \quad x < s_{n,i}, \\ 0, & \text{for} \quad x \geq s_{n,i}, \end{cases} \tag{4.23}$$

for the model with capacity and service constraints defined above.

Proof. The proof is similar to the proof for Theorem 2.8. □

4.4.3 The Infinite Horizon Model

The extension of the Markovian demand model with the lost sales assumption to the infinite horizon case can be obtained in the same way as the infinite horizon model presented in Chapter 2, Section 2.6. Particularly, let us assume that the cost functions and the density functions are independent of time, i.e.,

$$f_k(i,x) = f(i,x), \ q_k(i,x) = q(i,x), \ \varphi_{i,k} = \varphi_i,$$

and

$$c_k(i,x) = c(i,x) = K\mathbb{1}_{u>0} + c^i u, \ i \in \mathbb{I}, \ k = 0,1,2,\ldots.$$

It is important to observe that the fixed ordering cost K is independent of the state i as well, so that Assumption (4.13) holds. Thus, the dynamic programming equations become

$$\begin{aligned} v(i,x) &= f(i,x) + \inf_{u \geq 0} \{c(i,u) + \mathsf{E}[q(i, x+u-\xi^i) \\ &\quad + \alpha \sum_{j=1}^{L} p_{ij} v(j, (x+u-\xi^i)^+)]\}, \end{aligned} \tag{4.24}$$

where α is a given discount factor, $0 < \alpha \leq 1$, ξ^i is a random variable with density function $\varphi_i(\cdot)$, and $v(i,x)$ is the value function in any period with initial inventory x and demand state i. Therefore, the optimal inventory policies also become stationary, i.e., independent of period k.

THEOREM 4.5 *In the infinite horizon inventory problem with the lost sales assumption and the stationary cost and density functions, the optimal inventory control policy is given by*

$$\hat{u}(i,x) = \begin{cases} S_i - x, & \text{for} \quad x < s_i, \\ 0, & \text{for} \quad x \geq s_i. \end{cases} \tag{4.25}$$

Proof. The proof is similar to that of Theorem 2.5. □

The optimality of (s, S)-type policies for each of the above extended models can be established in a way similar to that in Chapter 2 with appropriate modifications for the lost sales situation. It should be noted that for the infinite horizon problem, the optimal policy parameters S_i and s_i are dependent only on the demand state i, and not on time.

4.5. Numerical Results

The purpose of this numerical study is twofold. First, by providing the algorithms and the computational results, we show that the theoretical results derived in the previous chapters are computationally feasible, and therefore, can be applied to solving real-world problems. Also, by comparing the performance of the results obtained from our models with that obtained based on the standard models reported in the literature, we demonstrate the advantages that can be achieved by Markovian demand models.

From a computational point of view, in order to obtain an optimal policy, it is necessary to solve the DP equations numerically. We have used continuous inventory state models throughout our analysis in the book. A discretization procedure will be necessary to make the computation tractable. Technical issues regarding the convergence of discretization procedures have been dealt with by Bertsekas (1976), for example. Under certain technical assumptions, a discretization procedure is shown to be stable in the sense that it yields suboptimal policies whose performance approaches the optimal value of the objective function arbitrarily closely as the discretization grids become finer and finer. Moreover, such discretization is always justified in practice since both demands and orders are in quantities of integer numbers. In particular, in a practical problem, the control space is always a compact set. This observation greatly simplifies the convergence problems associated with the discretization procedures.

We have established the optimality of an (s, S)-type policy for Markovian demand models with the lost sales assumption. Nevertheless, their computation is another matter. For the finite horizon models, we know that optimal policies are nonstationary even when all the system parameters (costs, density functions, etc.) are time-invariant. Computation has to be carried out for each period in the finite time horizon. Usually, this is done using a backward dynamic programming algorithm. Since computation in the finite horizon is rather straightforward, we will focus on the computation for the infinite horizon models.

It is clear that for infinite horizon problems, computation of the optimal policy is intractable since it would require an infinite pair of numbers $(s_n, S_n), n = 1, 2, \ldots,$ to be specified. Moreover, the computation of each (s_n, S_n) pair is based on the information about demands for the entire remaining infinite horizon. Therefore, we will only consider the computation for stationary infinite horizon models.

The algorithm we use for solving dynamic programming equations is essentially a modified value iteration algorithm. In the finite horizon case, the optimal policies can be computed in a finite number of iterations, which are equal to the number of periods of the problem. In the infinite horizon case with stationary data, it can be shown that the algorithm converges to an optimal policy; (see, e.g., Bertsekas (1976)).

Tables 4.1 and 4.2 list the numerical results obtained by the algorithm. For comparison purposes, we also compute the cost functions under a standard stationary (s, S) policy obtained without the consideration of the demand state variables as in Karlin and Fabens (1960), i.e, the decision space is restricted within the set of stationary (s, S) policies. In order to compute the stationary (s, S) policies, we first obtain the steady-state distribution of the demand state; hence, the limiting stationary demand distribution. The cost functions are then computed based on the stationary demand distribution. This approximating approach will enable one to obtain a stationary (s, S) policy. However, such a stationary (s, S) policy would be theoretically inferior to our state-dependent (s, S) policy as it is not optimal for the Markovian demand model. In fact, our numerical results confirm this assertion.

The results listed in Tables 4.1 and 4.2 show that a significant improvement is achieved by adopting a state-dependent (s, S) policy. A group of uniform distributions are used for the cases in Table 4.1, and truncated normal distributions are used in Table 4.2. The number of demand states $L = 3$ in all cases. The parameters c, K, α, h, and p are the unit purchase cost, the setup cost, the discount factor, the unit holding cost, and the unit shortage cost, respectively. The values $(s_i, S_i), i = 1, 2, 3$, are optimal parameters computed based on our Markovian model, and $(s, S)_{\mathrm{KF}}$ is the parameter of the stationary (s, S) policy obtained based on the model of Karlin and Fabens (KF).

Case	c	K	α	h	p	(s_1, S_1)	(s_2, S_2)	(s_3, S_3)	$(s, S)_{\text{KF}}$
1.1	0.50	1.0	0.90	1.00	5.00	7,9	13,15	18,20	14,16
1.2	0.50	1.5	0.90	1.00	5.00	6,9	12,15	17,20	13,17
1.3	1.00	2.0	0.90	1.00	5.00	5,8	11,14	16,19	12,16
1.4	0.50	2.0	0.90	1.00	2.00	3,8	9,14	14,19	8,14
1.5	0.50	2.0	0.90	1.00	1.00	0,0	4,11	9,16	1,10
1.6	0.50	2.0	0.95	1.00	2.00	3,8	9,14	14,19	8,13
1.7	0.50	2.0	0.85	1.00	2.00	3,8	9,14	14,19	8,14
1.8	1.00	2.0	0.90	1.00	2.00	0,0	6,11	11,16	5,12
1.9	1.00	5.0	0.90	1.00	2.00	0,0	3,11	8,16	2,13
1.10	2.00	5.0	0.90	1.00	2.00	0,0	0,0	0,0	0,0
1.11	2.50	5.0	0.90	1.00	2.00	0,0	0,0	0,0	0,0
1.12	1.00	5.0	0.90	2.00	4.00	2,8	8,14	13,19	7,14
1.13	1.00	5.0	0.90	1.00	4.00	2,8	8,14	13,19	9,16

Table 4.1. Numerical results for the lost sales model with uniform demand distribution.

Case	c	K	α	h	p	(s_1, S_1)	(s_2, S_2)	(s_3, S_3)	$(s, S)_{\text{KF}}$
2.1	0.50	1.0	0.90	1.00	5.00	11,14	15,18	16,19	14, 17
2.2	0.50	1.5	0.90	1.00	5.00	10,14	14,18	16,19	14, 17
2.3	1.00	2.0	0.90	1.00	5.00	9,12	13,16	15,18	13, 17
2.4	0.50	2.0	0.90	1.00	2.00	7,12	10,16	14,17	10, 15
2.5	0.50	2.0	0.90	1.00	1.00	2,9	5,13	10,16	4, 12
2.6	0.50	2.0	0.95	1.00	2.00	7,12	10,16	14,17	10, 15
2.7	0.50	2.0	0.85	1.00	2.00	7,12	10,16	14,17	10, 15
2.8	1.00	2.0	0.90	1.00	2.00	4,9	8,13	12,16	8, 14
2.9	1.00	5.0	0.90	1.00	2.00	1,9	4,13	9,16	4, 14
2.10	2.00	5.0	0.90	1.00	2.00	0,0	0,0	0,0	0, 0
2.11	2.50	5.0	0.90	1.00	2.00	0,0	0,0	0,0	0, 0
2.12	1.00	5.0	0.90	2.00	4.00	6,12	10,16	13,17	9, 15
2.13	1.00	5.0	0.90	1.00	4.00	6,12	10,16	13,17	10, 16

Table 4.2. Numerical results for the lost sales model with truncated normal demand distribution.

In Figures 4.1-4.8, we also plot the curves of the value functions for Cases 1.1-1.4 in Table 4.1 and Cases 2.1-2.4 in Table 4.2. The solid lines are the value functions $v(i, x)$ using the optimal policies (OP) based on the Markovian demand model, and the dotted lines are the corresponding value functions using a stationary (s, S) policy obtained according to the KF model. We observe from these figures that the state-dependent (s, S) policies perform better than the stationary (s, S) policies.

4.6. Concluding Remarks and Notes

The chapter, based on Cheng and Sethi (1999b) and Cheng (1996), extends the Markovian demand models to incorporate the case in which any unsatisfied demand is lost rather than backlogged. Our treatment of this case excludes some of the assumptions on the growth rate of the cost functions made in the literature. It is shown that like stochastic inventory models with backorders, the lost sales model with Markovian demands exhibits optimality of state-dependent (s, S) policies.

We also carry out a computational study to compare the performance of the optimal state-dependent (s, S) policies over the optimal standard (s, S) policies.

We have pointed out that the results for the lost sales case are obtained under the condition of zero ordering leadtime. With nonzero leadtimes, the optimality of (s, S)-type policies obtained in the backlog case do not hold for the lost sales case.

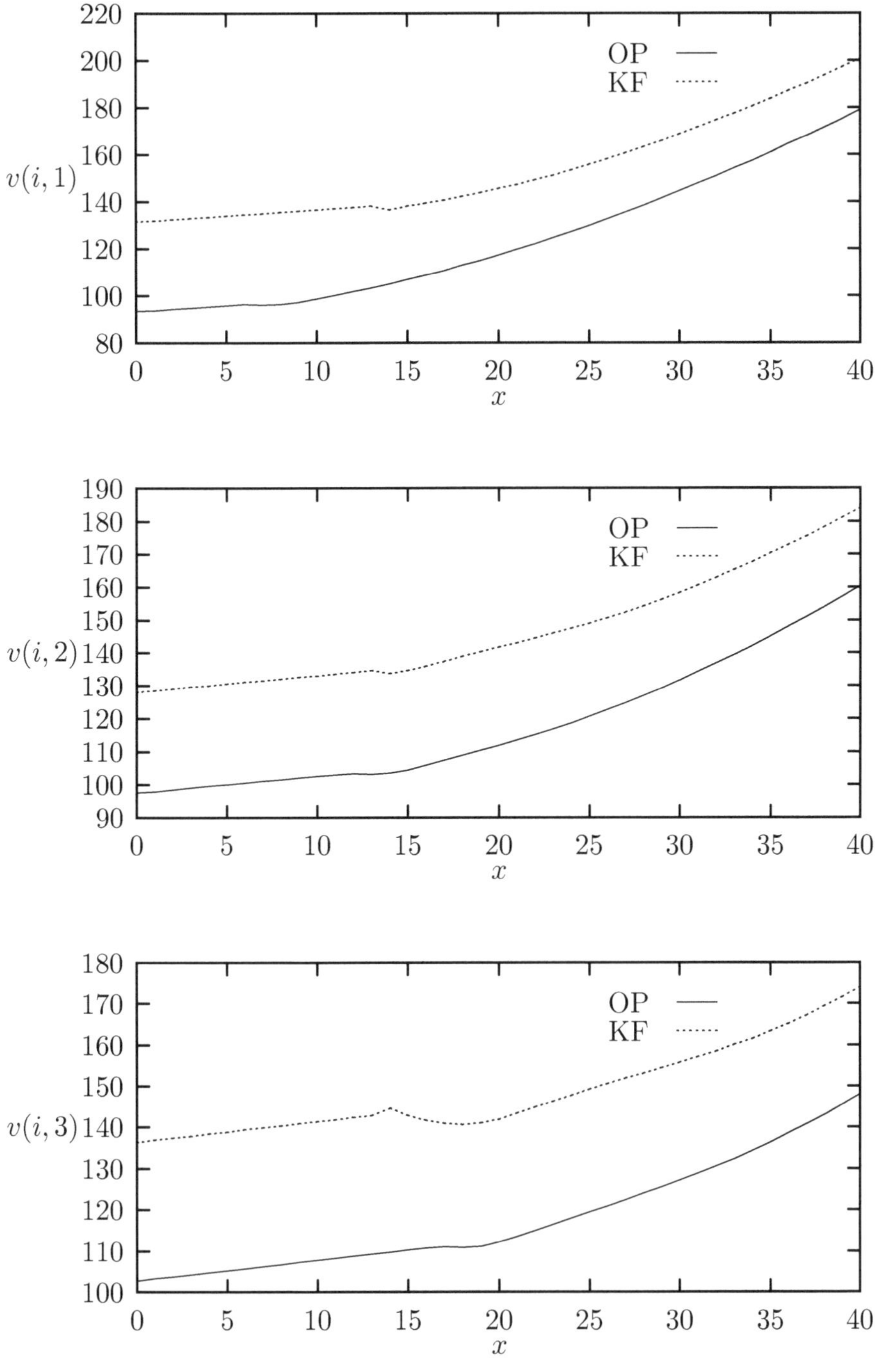

Figure 4.1. Results for Case 1.1.

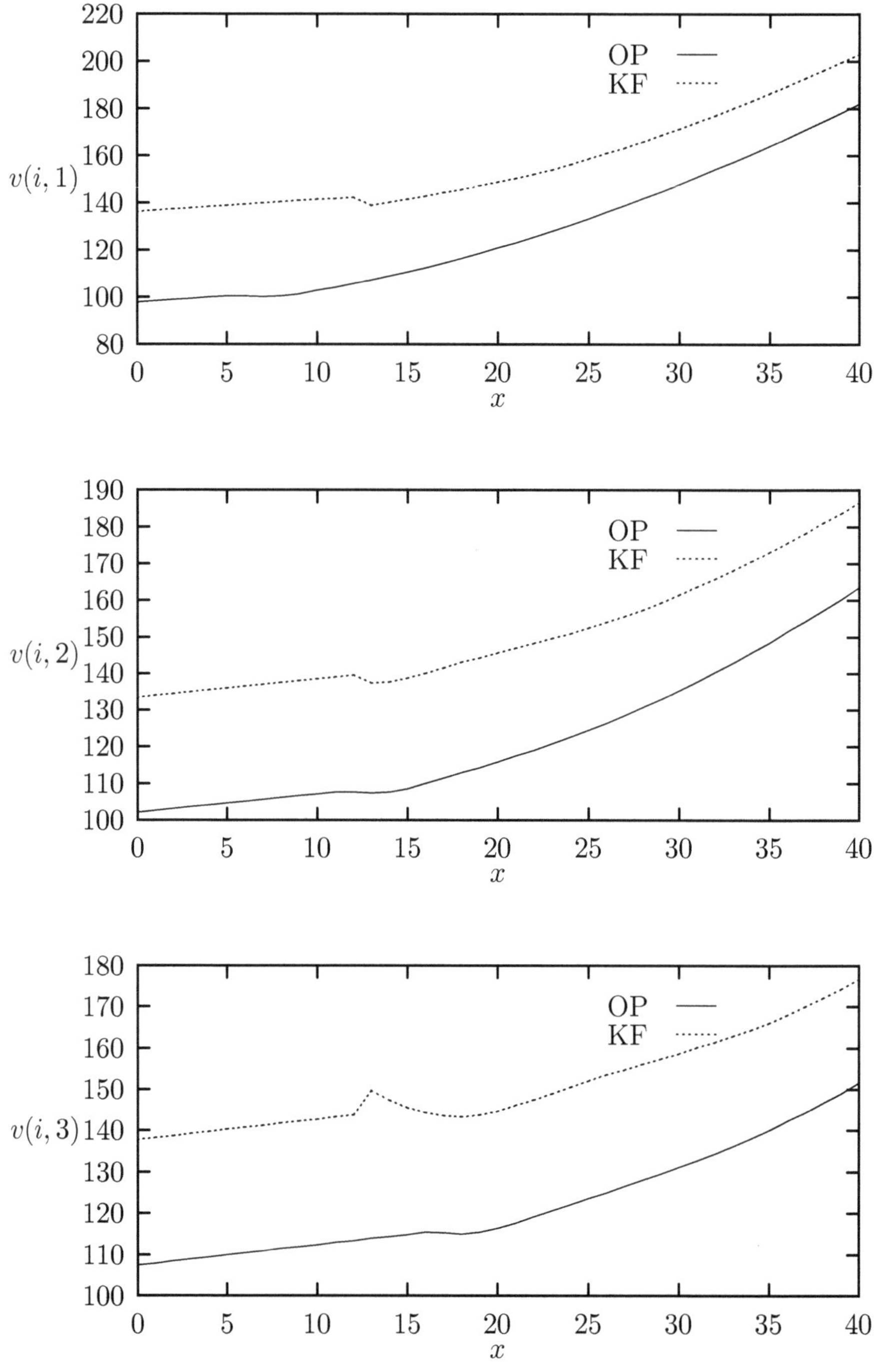

Figure 4.2. Results for Case 1.2.

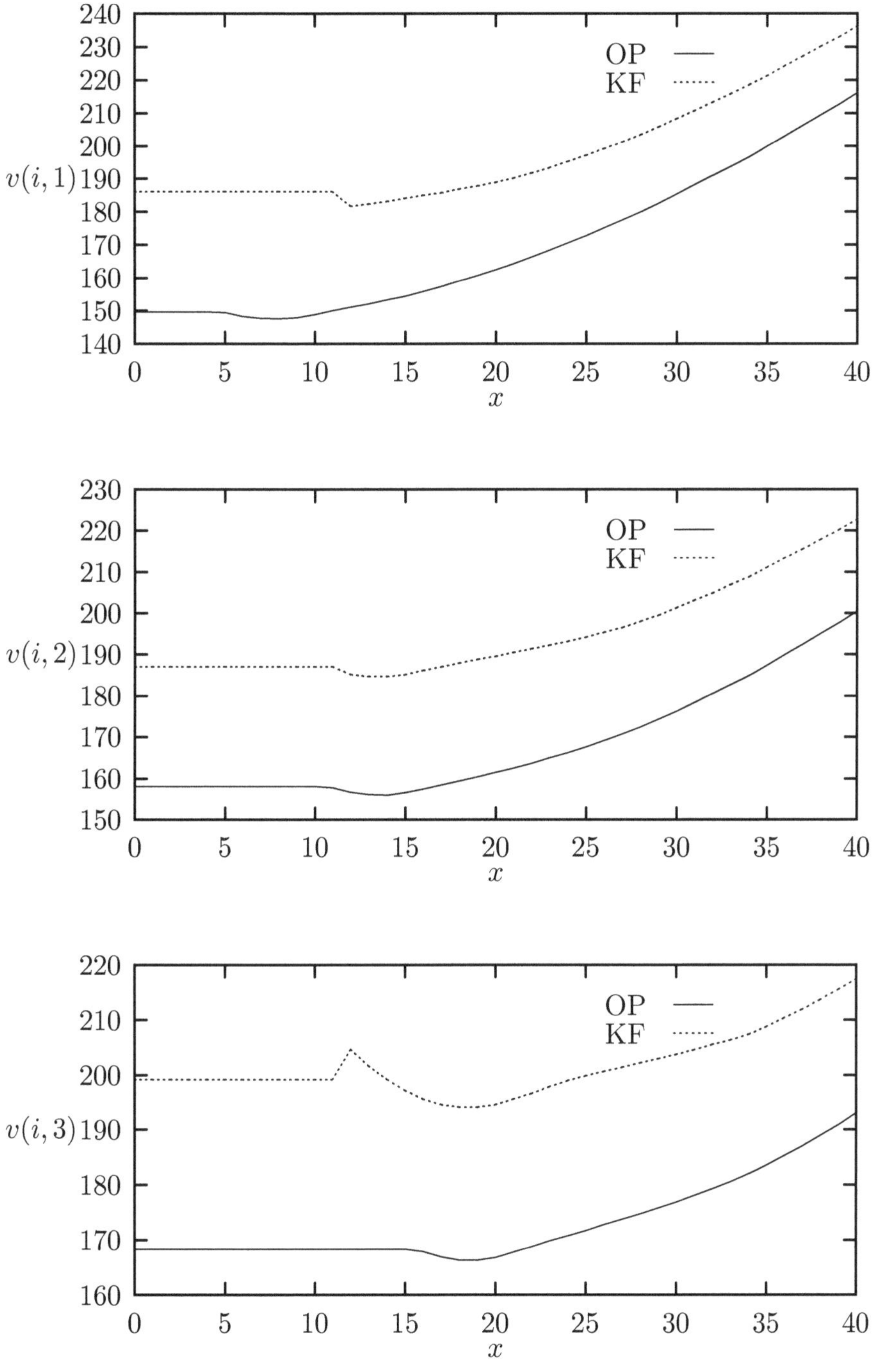

Figure 4.3. Results for Case 1.3.

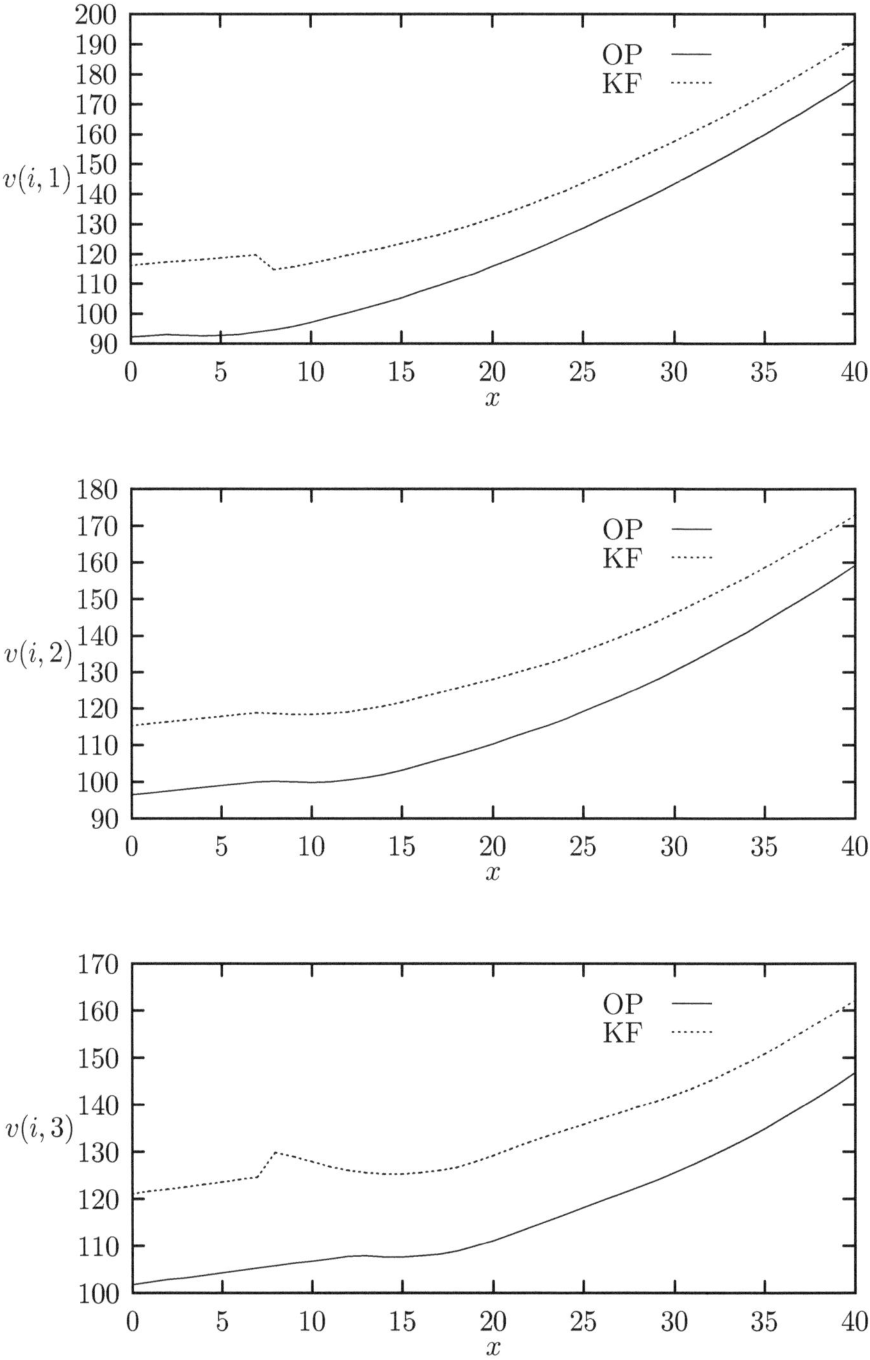

Figure 4.4. Results for Case 1.4.

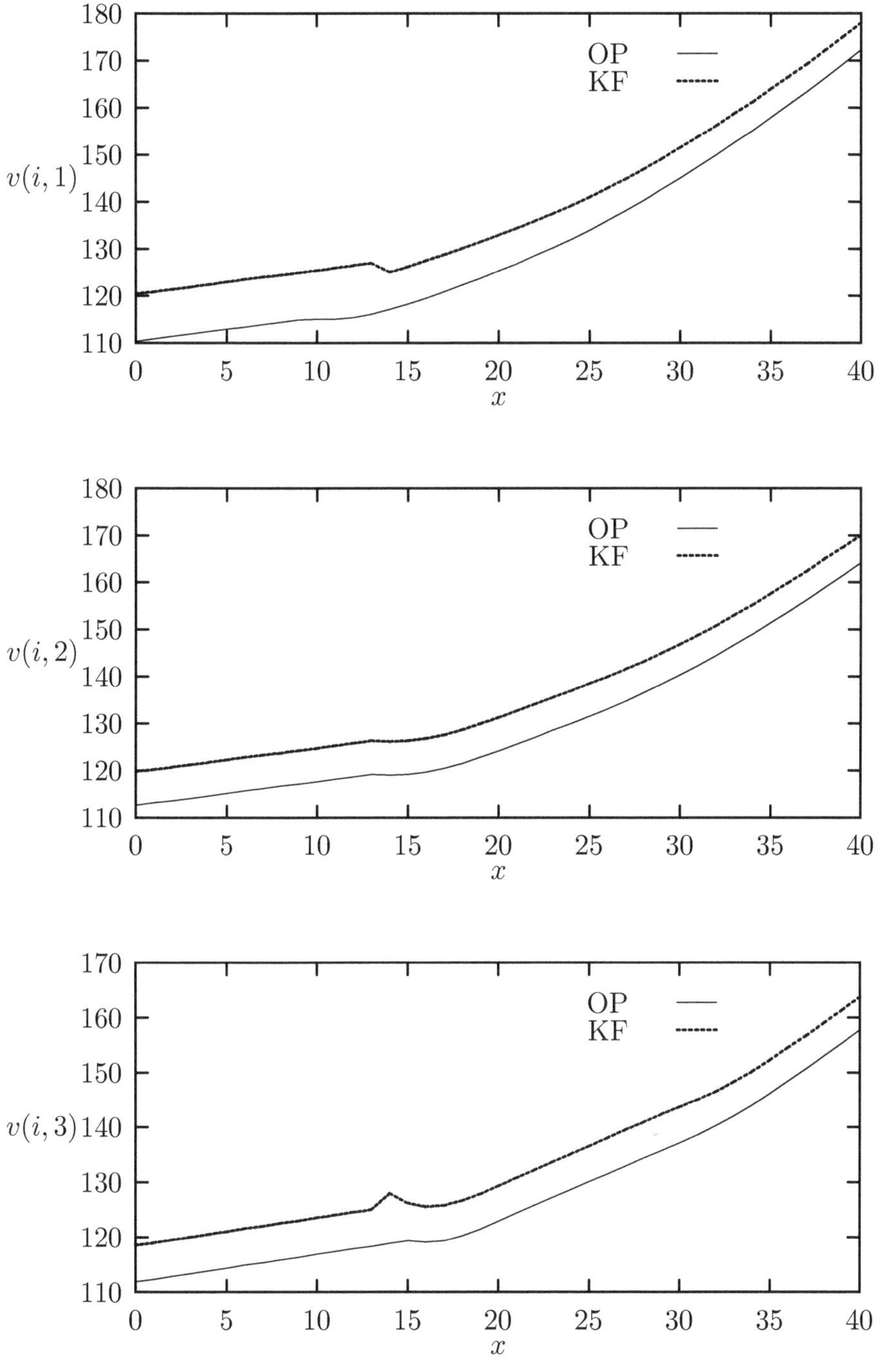

Figure 4.5. Results for Case 2.1.

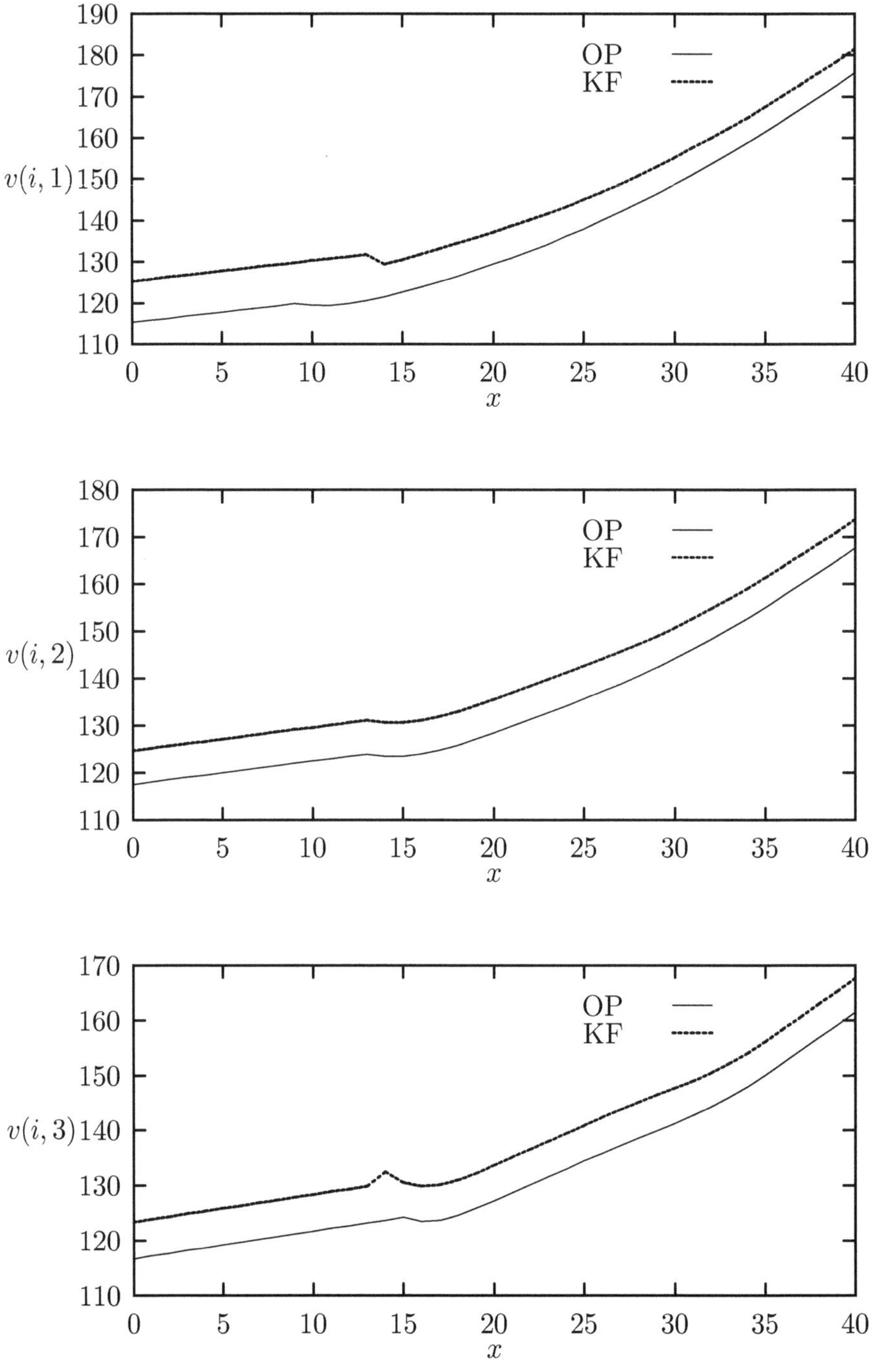

Figure 4.6. Results for Case 2.2.

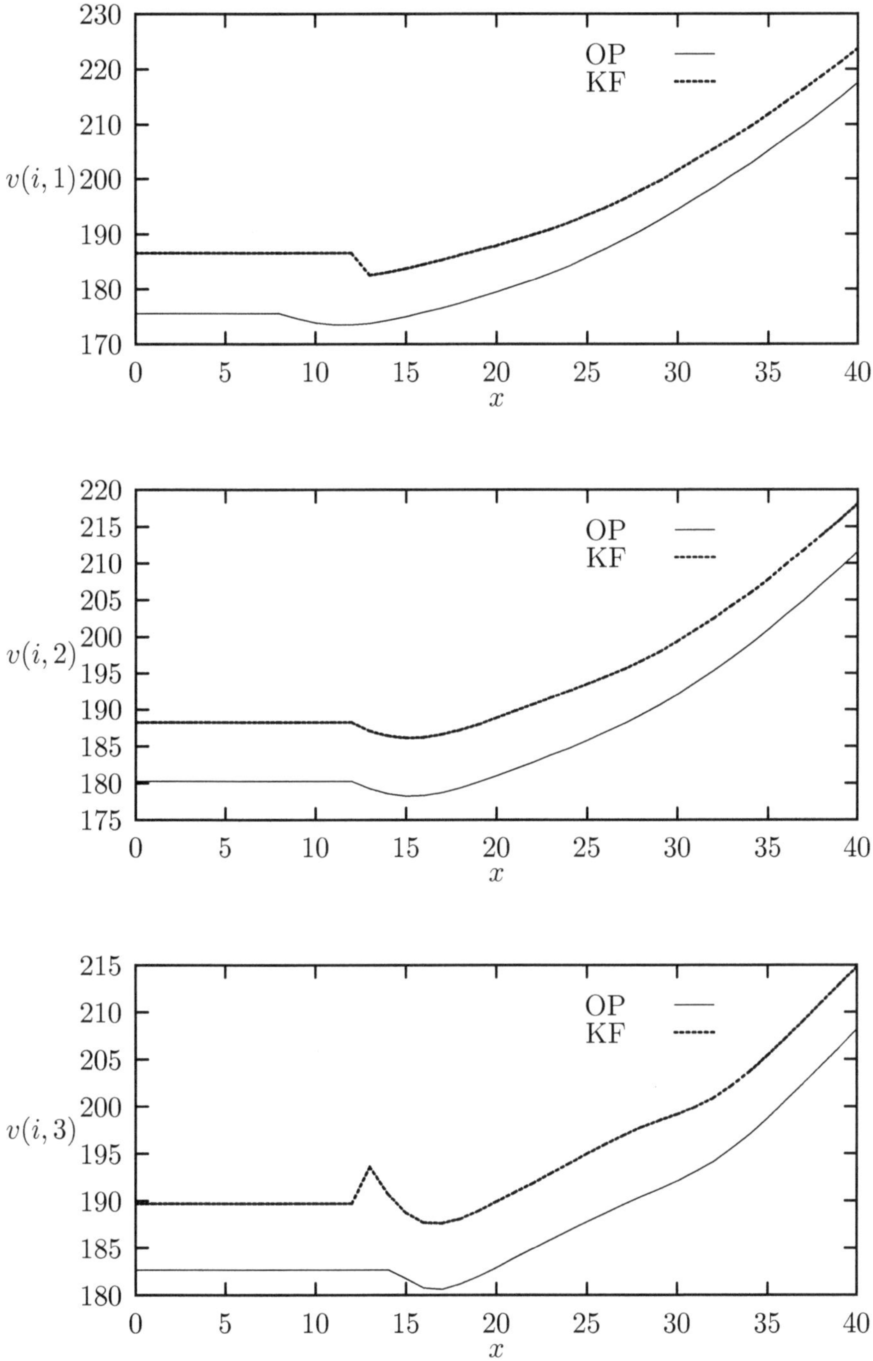

Figure 4.7. Results for Case 2.3.

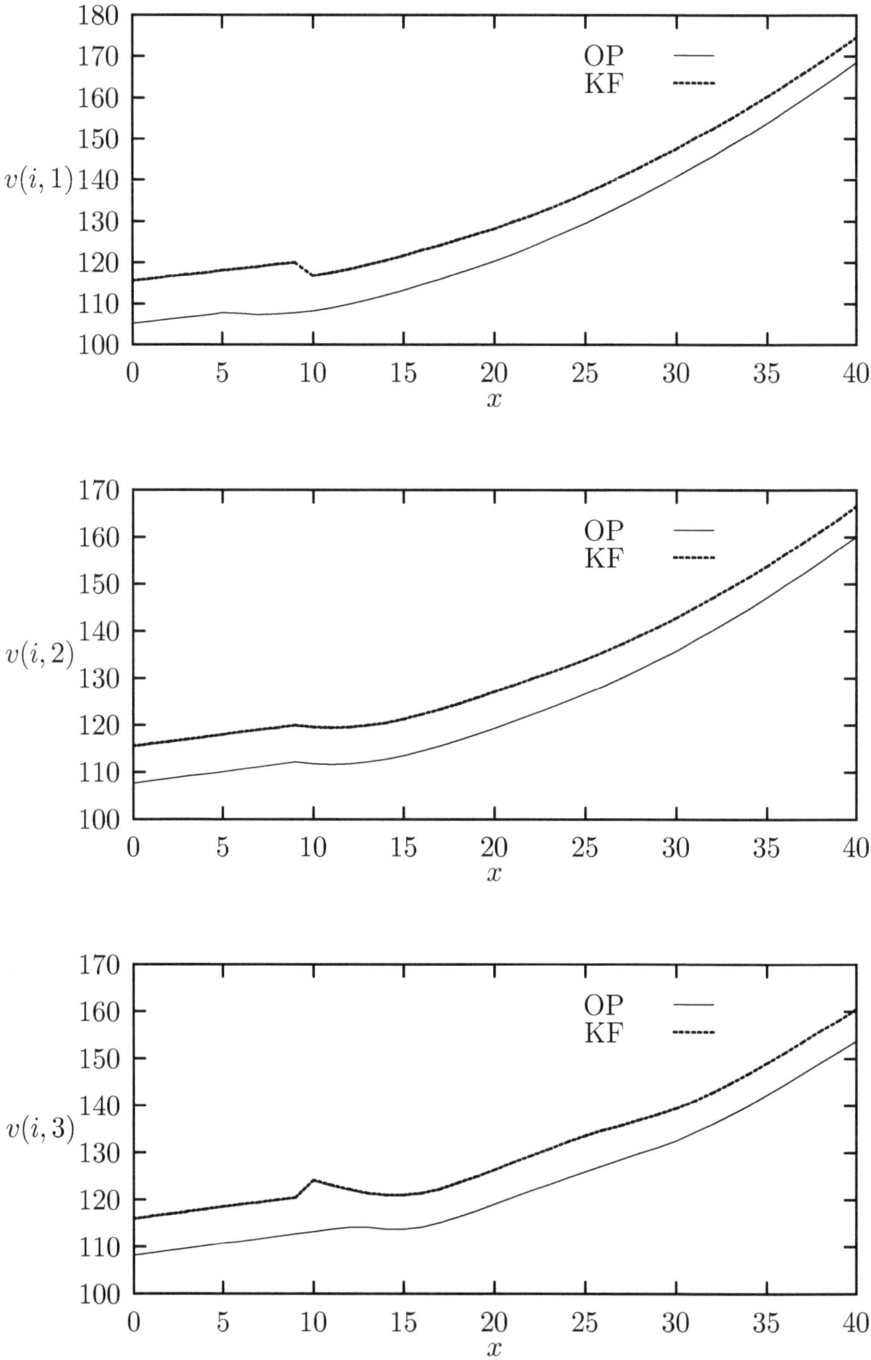

Figure 4.8. Results for Case 2.4.

III

AVERAGE COST MODELS

Chapter 5

AVERAGE COST MODELS WITH BACKORDERS

5.1. Introduction

This chapter is devoted to the study of a stochastic inventory problem with Markovian demand and fixed cost from the viewpoint of minimizing the long-run average cost of inventory/backlog and ordering. The purpose is to establish the dynamic programming equation or average cost optimality equation for the problem, prove the existence of an optimal feedback (or Markov) policy, and show that a feedback policy of (s, S)-type is optimal.

A Markov chain, whose states represent possible states of the environment or the economy, underlies the Markovian demand process in the sense that the distribution of the demand in any given period depends on the environmental state in that period. Furthermore, these states may also affect various cost parameters in the model. Problems with Markovian demand and fixed costs were considered in Part II, but only with the discounted cost criterion. There, we rigorously established the existence of an optimal state-dependent (s, S) policy. Results from Chapter 2 that we need for our analysis of the average cost problems are recapitulated in Section 5.3.

The problems of long-run average cost minimization are mathematically much harder than those with discounted costs; (see Arapostathis *et al.* (1993) for a survey of discrete-time average cost problems). In the context of inventory problems, Iglehart (1963b) and Veinott and Wagner (1965) were the first to study the issue of the existence of an optimal (s, S) policy for average cost problems with independent demands, linear holding and backlog costs, and ordering costs consisting of a fixed cost and a proportional variable cost. Iglehart obtained the

D. Beyer et al., *Markovian Demand Inventory Models*, International Series in Operations Research & Management Science 108,
DOI 10.1007/978-0-387-71604-6_5, © Springer Science+Business Media, LLC 2010

stationary distribution of the inventory/backlog (or surplus) level given an (s, S) policy using renewal theory arguments (see also Karlin (1958a) and Karlin (1958b)), and developed an explicit formula for the stationary average cost $\mathcal{L}(s, S)$, $s \leq S$, associated with the policy. His subsequent analysis is carried out under the assumption that $\mathcal{L}(s, S)$ is continuously differentiable and that there exists a pair (s^*, S^*), $s^* < S^*$, which minimizes $\mathcal{L}(s, S)$ and satisfies the first-order conditions for a local minimum. While he does not specify these assumptions explicitly, he uses them in showing the key result that the minimum average cost of finite horizon problems approaches $\mathcal{L}(s^*, S^*)$ asymptotically as the horizon becomes infinitely large.

Veinott and Wagner deal with the case of discrete demands. With an additional argument suggested by Derman (1965), they were able to convert the Iglehart result, under the same assumptions on $\mathcal{L}(s, S)$, into the optimality of the (s^*, S^*) policy; (see also Veinott (1966)). It should be noted that Veinott and Wagner, dealing with discrete demands, require a discrete demand version of the results in Iglehart.

A great deal of research has been carried out in connection with the average cost (s, S) models since then. With the exception of Zheng (1991) and Huh *et al.* (2008), most of the research is concerned with the computation of (s^*, S^*) that minimizes $\mathcal{L}(s, S)$, and not with the issue of establishing the optimality of an (s, S)-type policy. The examples are Stidham (1977), Zheng and Federgruen (1991), Federgruen and Zipkin (1984), Hu *et al.* (1993), and Fu (1994). For other references, the reader is directed to Porteus (1985), Sahin (1990), and Zheng and Federgruen (1991). Furthermore, this literature has assumed that taken together, the papers of Iglehart (1963b) and Veinott and Wagner (1965) have established the optimality of an (s, S) policy for the problem. This is not quite the case, however, since the assumptions on $\mathcal{L}(s, S)$ implicit in Iglehart (1963b), to our knowledge, have not been satisfactorily verified. In Chapter 9 based on Beyer and Sethi (1999), we discuss these issues in detail, as well as derive the required results that are missing in Iglehart (1963b).

Zheng (1991) has provided a rigorous proof of the optimality of an (s, S)-type policy in the case of discrete demands. He was able to use the theory of countable state Markov decision processes in the case when the solution of the average cost optimality equation for the given problem is bounded, which is clearly not the case since the inventory cost is unbounded. So, he relaxed the problem by allowing inventory disposals, and since the inventory costs are charged on ending inventories in his problem (see Remark 5.3), he obtained a bounded solution for the average cost optimality equation of the relaxed problem, which involves a dispose-down-to-S component. But the dispose-down-to-S component of the optimal policy would be invoked in the relaxed problem only in the first period (and only when the initial inventory is larger than S), which has no influence on the long-run average cost of the policy. It therefore follows that the (s, S) policy, without the dispose-down-to-S component, will also be optimal for the original problem.

More importantly for the purpose of this chapter, the methodology of Iglehart (1963b) and Veinott and Wagner (1965) crucially depends on being able to obtain the stationary distribution of the surplus variable, and thus on the specifics of the problem. For more general problems such as the one involving Markovian demand considered in this chapter, where it may not be possible to explicitly obtain the stationary distribution of the state variables of the system, a general methodology is required. Also, the disposal device used by Zheng (1991) does not work in the case of Markovian demand. Furthermore, without establishing a bounded solution of the average cost optimality equation, the optimality conditions established in the theory of Markov decision processes are, in Zheng's own words, tedious to verify.

The problem under consideration in this chapter involves a Borel state space, a noncompact action set, and unbounded costs, the study of which has begun only recently and is far from complete. The general results obtained to date involve restrictive and difficult to verify conditions; (see Arapostathis *et al.* (1993)). Thus, it is important to study special problems of this kind. Sethi *et al.* (2005a) have used the vanishing discount approach for optimization of continuous-time stochastic manufacturing systems with unreliable machines. They require the concept of viscosity solutions to the average cost optimality equation because of the continuous-time nature of their model.

In this chapter, we use a vanishing discount approach to study the long-run average cost problem with Markovian demand, fixed ordering cost, and convex surplus cost. The idea is to use the results obtained in Chapter 2 for the discounted cost infinite horizon problem and analyze them as the discount factor approaches one. In order to use this procedure, we impose an additional condition for the demands to be uniformly bounded. The unique features of our average cost problem are the Markovian demand and the fixed ordering cost. Presence of the latter renders the value functions involved to be nonconvex, and thus harder to handle as also mentioned in Iglehart (1963b).

The plan of this chapter is as follows. In the next section, we provide a precise formulation of the problem. Relevant results for the discounted cost problem obtained in Chapter 2 are summarized in Section 5.3. The asymptotic behavior of the differential discounted value function, as the discount rate approaches zero, is obtained in Section 5.4. In Section 5.5, we develop the vanishing discount approach to establish the average cost optimality equation. The associated verification theorem is proved in Section 5.6, and the theorem is used to show that a state-dependent (s, S) policy is optimal for the problem. Section 5.7 concludes the chapter with suggestions for future research.

5.2. Formulation of the Model

In order to specify the stationary, discrete-time, infinite horizon inventory problem under consideration, we introduce the following notation and basic assumptions.

$(\Omega, \mathcal{F}, \mathsf{P})$ = the probability space;

$\mathbb{I}$ = $\{1, 2, \ldots, L\}$, finite collection of possible demand states;

i_k = the demand state in period k, $k \in \mathbb{Z} = \{0, 1, 2, \ldots\}$;

$\{i_k\}$ = a Markov chain with the $(L \times L)$-transition matrix $P = \{p_{ij}\}$;

ξ_k = the demand in period k; it depends on i_k but not on k, and is independent of past demand states and past demands;

$\varphi_i(\cdot)$ = the conditional density function of ξ_k when $i_k = i$;

$\Phi_i(\cdot)$ = the distribution function corresponding to φ_i;

u_k = the nonnegative order quantity in period k;

x_k = the surplus (inventory/backlog) level at the beginning of period k (or, at the end of period $k-1$);

$c(i, u)$ = the cost of ordering $u \geq 0$ units in period k when $i_k = i$;

$f(i, x)$ = the surplus cost when $i_k = i$ and $x_k = x$; $f_k(i, 0) \equiv 0$.

We suppose that orders are placed at the beginning of a period, delivered instantaneously, and followed by the period's demand; (see Figure 2.1 in Chapter 2). Unsatisfied demands are fully backlogged.

In what follows, we list the assumptions that are needed to derive the main results of the chapter in Sections 5.5 and 5.6. Because not all the results proved in this chapter require all of these assumptions, we label them as follows for ease of specifying the assumptions required in the statements of the specific results proved in this chapter.

(i) The production cost is given by $c(i,u) = K\mathbb{I}_{u>0} + c_i u$, where the fixed ordering cost is denoted by $K \geq 0$ and the variable cost by $c_i \geq 0$.

(ii) For each i, the surplus cost function $f(i,\cdot)$ is convex and asymptotically linear, i.e., $f(i,x) \leq C(1+|x|)$ for some $C > 0$. Also, $f(i,0) = 0$.

(iii) There is a state $l \in \mathbb{I}$ such that $f(l,x)$ is not identically zero for $x \leq 0$.

(iv) There is a state $g \in \mathbb{I}$ such that $f(g,x)$ is not identically zero for $x \geq 0$.

(v) The production and inventory costs satisfy

$$c_i x + \sum_{j=1}^{L} p_{ij} \int_0^\infty f(j, x-z) d\Phi_i(z) \to \infty \text{ as } x \to \infty. \tag{5.1}$$

(vi) The Markov chain $(i_k)_{k=0}^\infty$ is irreducible.

(vii) There is a state $h \in \mathbb{I}$ such that $1 - \Phi_h(\varepsilon) = \rho > 0$ for some $\varepsilon > 0$.

(viii) There is an M, $0 < M < \infty$, such that $0 \leq \xi_k \leq M$, a.s.

REMARK 5.1 Assumptions (i) and (ii) reflect the usual structure of the production and inventory costs to prove the optimality of an (s_i, S_i) policy. Note that K is the same for all i. In the stationary case, this is equivalent to the condition (2.18) required in the nonstationary model for the existence of an optimal (s_i, S_i) policy; (see Chapter 2). Assumptions (iii) and (iv) rule out trivial cases which lead to degenerate optimal policies. In fact, if Assumption (iii) is violated, the optimal policy is never to order. If Assumption (iii) holds and Assumption (iv) is violated, it is optimal to wait for a period with a demand state for which c_i is minimal and then to order an infinite amount. Assumption (v) means that either the unit ordering cost $c_i > 0$ or the second term in (5.1), which is the expected holding cost, or both, go to infinity as the surplus level x goes to infinity. While related, Assumption (v) neither implies nor is implied by Assumption (iv). Assumption (v) is borne out of practical considerations and is not very restrictive. In addition, it rules out unrealistic trivial cases such as the one with $c_i = 0$ and $f(i,x) = 0, x \geq 0$, for each i, which implies ordering an infinite amount whenever an order is placed. Assumptions (iv) and (v) generalize the usual assumption made by Scarf (1960) and others, that the unit inventory holding cost $h > 0$.

REMARK 5.2 Assumptions (vi) and (vii) are needed to deplete any given initial inventory in a finite expected time. While Assumption (vii) says that in at least one state h, the expected demand is strictly larger than zero, Assumption (vi) implies that the state h would occur infinitely often with finite expected intervals between successive occurrences. Assumption (viii) requires that the demand to be bounded is realistic. At any rate, it is not a crucial assumption, since it could be replaced by a growth condition on the distribution functions. Moreover, it does simplify the proofs considerably.

The objective is to minimize the expected long-run average cost

$$J(i,x;U) = \limsup_{N\to\infty} \frac{1}{N} \mathsf{E}\left\{\sum_{k=0}^{N-1} [c(i_k,u_k) + f(i_k,x_k)]\right\}, \tag{5.2}$$

with $i_0 = i$ and $x_0 = x$, where $U = (u_0, u_1, \ldots)$, $u_i \geq 0$, $i = 0, 1, \ldots$, is a *history-dependent* or *nonanticipative* decision (order quantities) for the problem. Such a control U is termed admissible. Let $\mathcal{U}$ denote the class of all admissible controls. The surplus balance equations are given by

$$x_{k+1} = x_k + u_k - \xi_k,\ k = 0, 1, \ldots. \tag{5.3}$$

Our aim is to show that there exist a constant λ^* termed the optimal average cost, which is independent of the initial i and x, and a control $U^* \in \mathcal{U}$ such that

$$\lambda^* = J(i,x;U^*) \leq J(i,x;U), \quad \text{for all} \quad U \in \mathcal{U}, \tag{5.4}$$

and

$$\lambda^* = \lim_{N\to\infty} \frac{1}{N} \mathsf{E}\left\{\sum_{k=0}^{N-1} [c(i_k,u_k^*) + f(i_k,x_k^*)]\right\}, \tag{5.5}$$

where x_k^*, $k \in \mathbb{Z}$, is the surplus process corresponding to U^* with $i_0 = i$ and $x_0 = x$.

To prove these results we will use the vanishing discount approach. That is, by letting the discount factor α in the discounted cost problem approach one, we will show that we can derive a dynamic programming equation whose solution provides an average optimal control and the associated minimum average cost λ^*.

For this purpose, we recapitulate relevant results for the discounted cost problem obtained in Chapter 2.

REMARK 5.3 Note that the objective function (5.2) is slightly, but not essentially, different from that used in the classical literature. Whereas we base the surplus cost on the initial surplus in each period, the usual practice in the literature is to charge the cost on the ending surplus levels, which means to have $f(i_k, x_{k+1})$ instead of $f(i_k, x_k)$ in (5.2). Note that x_{k+1} is also the ending inventory in period k. It should be obvious that this difference in the objective functions does not change the long-run average cost for any admissible policy.

5.3. Discounted Cost Model Results from Chapter 2

Consider the model formulated above with the average cost objective (5.2) replaced by the extended real-valued objective function

$$J^{\alpha}(i,x;U) = \sum_{k=0}^{\infty} \alpha^k \mathsf{E}[c(i_k,u_k) + f(i_k,x_k)], \quad 0 \le \alpha < 1. \tag{5.6}$$

Define the value function with $i_0 = i$ and $x_0 = x$ as

$$v^{\alpha}(i,x) = \inf_{U \in \mathcal{U}} J^{\alpha}(i,x;U). \tag{5.7}$$

Let $\mathbb{B}_0$ denote the class of all continuous functions from $\mathbb{I} \times \mathbb{R}$ into $[0,\infty)$, and the pointwise limits of sequences of these functions; (see Feller (1971)). Note that it includes piecewise-continuous functions. Let $\mathbb{B}_1$ denote the space of functions in $\mathbb{B}_0$ that are of linear growth, i.e., for any $b \in \mathbb{B}_1$, $0 \le b(i,x) \le C_b(1+|x|)$ for some $C_b > 0$. Let $\mathbb{B}_2$ denote the subspace of functions in $\mathbb{B}_1$ that are uniformly continuous with respect to $x \in R$. For any $b \in \mathbb{B}_1$, we define the notation

$$Fb(i,y) = \sum_{j=1}^{L} p_{ij} \int_0^M b(j, y-z) d\Phi_i(z). \tag{5.8}$$

The following results for the more general case of nonstationary cost functions are proved in Chapter 2 for $\alpha > 0$; (see Theorems 2.4 and 2.5).

THEOREM 5.1 *Let Assumptions* (i), (ii), (v) *and* (vi) *hold. Then, we have the following.*

(a) *The value function $v^{\alpha}(\cdot,\cdot)$ is in $\mathbb{B}_2$ and is the unique solution of the dynamic programming equation*

$$v^{\alpha}(i,x) = f(i,x) + \inf_{u \ge 0}\{c(i,u) + \alpha F(v^{\alpha})(i,x+u)\}. \tag{5.9}$$

(b) *$v^{\alpha}(i,\cdot)$ is K-convex and there are real numbers $(s_i^{\alpha}, S_i^{\alpha})$, $s_i^{\alpha} \le S_i^{\alpha}$, $i \in \mathbb{I}$, such that the feedback policy $\hat{u}_k^{\alpha}(i,x) = (S_i^{\alpha} - x)\mathbb{1}_{x < s_i^{\alpha}}$ is optimal.*

Hereafter, we will omit the additional superscript α on the control policies for ease of notation. Thus, for example, $\hat{u}_k^\alpha(i,x)$ will be denoted simply as $\hat{u}_k(i,x)$. Since we do not consider the limits of the control variables as $\alpha \to 1$, the practice of omitting the superscript α will not cause any confusion. In any case, the dependence of controls on α will always be clear from the context.

5.4. Limiting Behavior as the Discount Factor Approaches 1

To insure a "smooth" limiting behavior for $\alpha \to 1$, we prove in Lemma 5.3 that $v^\alpha(i,\cdot)$ is locally equi-Lipschitzian. To do so, some preliminary results are needed.

For any $y > 0$, define

$$\tau_y := \min\{n : \sum_{k=0}^{n} \xi_k \geq y\} \tag{5.10}$$

to be the first index for which the cumulative demand is not less than y.

LEMMA 5.1 *Let Assumptions* (vi) *and* (vii) *hold. Then,* $\mathsf{E}_l(\tau_y) < \infty$, *where*

$$\mathsf{E}_l(\eta) = \mathsf{E}(\eta | i_0 = l). \tag{5.11}$$

Proof. We first show that $\mathsf{E}_l(\tau_\varepsilon) < \infty$ for $\varepsilon > 0$ defined in Assumption (vii) and any l. Let $\psi_n := \sum_{k=0}^{n} \xi_k$. Clearly, τ_ε is a nonnegative integer random variable, and therefore it is

$$\mathsf{E}_l(\tau_\varepsilon) = \sum_{k=0}^{\infty} \mathsf{P}(\tau_\varepsilon > k | i_0 = l) = \sum_{k=0}^{\infty} \mathsf{P}(\psi_k < \varepsilon \,|\, i_0 = l).$$

In view of Assumption (vi), we can apply Theorem B.6.3 and see that there exists a stationary distribution $\nu = (\nu_1, \nu_2, \ldots, \nu_L)$ with $\nu_i > 0$, $i = 1, 2, \ldots, L$. From Theorem B.6.4, we see that there is an N such that

$$\mathsf{P}(i_k = i \text{ for some } k \in \langle 0, N{-}1\rangle \,|\, i_0 = l) \geq \nu_i/2, \quad i = 1, 2, \ldots, L. \tag{5.12}$$

Define $\psi^a := \sum_{k=aN}^{(a+1)N-1} \xi_k$ for any nonnegative integer a. By (5.12) and Assumption (vii), we have

$$\begin{aligned}\mathsf{P}(\psi^j \geq \varepsilon \,|\, i_0 = l) &\geq \mathsf{P}(\xi_k \geq \varepsilon \text{ and } i_{k-1} = h \text{ for some} \\ &\qquad\qquad k \in \langle jN, (j+1)N{-}1\rangle \,|\, i_0 = l) \\ &\geq \rho\nu_h/2.\end{aligned}$$

Thus,

$$\mathsf{P}(\psi^j < \varepsilon \,|\, i_0 = l) = 1 - \mathsf{P}(\psi^j \geq \varepsilon \,|\, i_0 = l) \leq 1 - \rho\nu_h/2. \qquad (5.13)$$

Since the set $\{\psi_{aN} < \varepsilon\} \subset \{\psi^0 < \varepsilon, \psi^1 < \varepsilon, \ldots, \psi^a < \varepsilon\}$, we have

$$\begin{aligned} \mathsf{P}(\psi_{aN} < \varepsilon \,|\, i_0 = l) &\leq \mathsf{P}(\psi^0 < \varepsilon, \psi^1 < \varepsilon, \ldots, \psi^a < \varepsilon \,|\, i_0 = l) \\ &\leq (1 - \rho\nu_h/2)^a. \end{aligned}$$

From the decreasing property of $\mathsf{P}(\psi_k < \varepsilon \,|\, i_0 = l)$ with respect to k, we obtain

$$\begin{aligned} \mathsf{E}_l(\tau_\varepsilon) &= \sum_{a=0}^{\infty} \sum_{k=aN}^{(a+1)N-1} \mathsf{P}(\psi_k < \varepsilon \,|\, i_0 = l) \\ &\leq \sum_{a=0}^{\infty} \sum_{k=aN}^{(a+1)N-1} \mathsf{P}(\psi_{aN} < \varepsilon \,|\, i_0 = l) \\ &\leq N \sum_{a=0}^{\infty} (1 - \rho\nu_h/2)^a = 2N/\rho\nu_h < \infty. \end{aligned}$$

To prove the finiteness of $\mathsf{E}_l(\tau_y)$, we first show that $\mathsf{E}_l(\tau_{n\varepsilon}) < \infty$ for each $l \in \mathbb{I}$. We have already proved this for $n = 1$. Assume that it is true for $n = k$, i.e., $\mathsf{E}_l(\tau_{k\varepsilon}) < \infty$ for each $l \in \mathbb{I}$. Then we have

$$\mathsf{E}_l(\tau_{(k+1)\varepsilon}) \leq \mathsf{E}_l(\tau_{k\varepsilon}) + \max_{j \in \mathbb{I}}\{\mathsf{E}_j(\tau_\varepsilon)\} < \infty,$$

and, therefore, by induction $\mathsf{E}_l(\tau_{n\varepsilon}) < \infty$. Obviously, $\mathsf{E}_l(\tau_y)$ is increasing in y. Choosing $m \geq y/\varepsilon$, we finally obtain $\mathsf{E}_l(\tau_y) \leq \mathsf{E}_l(\tau_{m\varepsilon}) < \infty$. □

If the ordering cost function in the demand state i has the form $c(i, u) = K\mathbb{1}_{u>0} + c_i u$, we know from Chapter 2 that in the stationary case, an (s_i, S_i) policy is optimal.

LEMMA 5.2 *Under Assumptions* (i)-(iii) *and* (v)-(viii), *there are real numbers* $0 < \alpha_0 < 1$ *and* $C_2 > -\infty$ *such that* $\max_{i \in \mathbb{I}}\{s_i^\alpha\} \geq C_2$ *for all* $1 > \alpha \geq \alpha_0$.

Proof. For any fixed discount factor $0 < \alpha < 1$, let U be an optimal strategy with parameters (s_i^α, S_i^α). Because of Assumptions (ii) and (iii), there are constants $0 \geq C_3 > -\infty$ and $q > 0$, such that $f(l, C_3 - y) \geq qy$ for all $y > 0$. Let us fix a positive integer T and let $C_2 = C_3 - MT$. Assume $s_i^\alpha < C_2$ for each $i \in \mathbb{I}$. In what follows, we specify a value of T, namely T^*, in terms of which we will construct an alternative strategy that is better than U, at least when the initial surplus $x = C_3$.

Consider the alternate policy $\tilde{U} = (\tilde{u}_0, \tilde{u}_1, \ldots)$, defined by

$$\tilde{u}_0 = 0, \;\; \tilde{u}_k = \xi_{k-1} \text{ for } k = 1, 2, \ldots, T,$$

and

$$\tilde{u}_k = [x_k + u_k - \tilde{x}_k]^+ \text{ for } k \geq T+1.$$

For convenience of notation we define $\xi_{-1} = 0$.

We investigate the strategy $\tilde{U}$ with the initial surplus $\tilde{x}_0 = C_3$. Let $\tilde{x}_k$ denote the inventory level in period k resulting from $\tilde{U}$. With the initial inventory level $x_0 = \tilde{x}_0 = C_3$, we know that under the optimal policy U, no order is placed in or before period T, since the corresponding surplus $x_k \geq C_3 - Mk \geq C_2 > s_i^\alpha$, $k = 0, 1, \ldots, T$. It is easy to see that $\tilde{x}_k = C_3 - \xi_{k-1} \geq x_k = C_3 - \sum_{t=0}^{k-1} \xi_t$, $k \leq T$. For $k \geq T+1$, we have the following situation. As long as the inventory level after ordering for U is less than the inventory level before ordering for $\tilde{U}$, the strategy $\tilde{U}$ orders nothing. As soon as $x_k + u_k$ is greater than or equal to $\tilde{x}_k$, we order up to the level $x_k + u_k$, and the inventory level is the same for both strategies from then on. Obviously, the expected cost corresponding to U from period $T+1$ on is not smaller than the one corresponding to $\tilde{U}$ from $T+1$ on. Therefore,

$$\begin{aligned} & v^\alpha(i, C_3) - J^\alpha(i, C_3, \tilde{U}) \\ &= \mathsf{E}\left(\sum_{k=0}^{\infty} \alpha^k (c(i_k, u_k) + f(i_k, x_k) - c(i_k, \tilde{u}_k) - f(i_k, \tilde{x}_k)) \right) \qquad (5.14) \\ &\geq \mathsf{E}\left(\sum_{k=0}^{T} \alpha^k (f(i_k, C_3 - \sum_{t=0}^{k-1} \xi_t) - c(i_k, \xi_{k-1}) - f(i_k, C_3 - \xi_{k-1})) \right). \end{aligned}$$

Furthermore, because of the boundedness of the demand, we know that there is a constant C_4 such that $\mathsf{E}(c(i_k, \xi_{k-1}) + f(i_k, C_3 - \xi_{k-1})) \leq C_4$ for all k and, therefore,

$$\mathsf{E}\left(\sum_{k=0}^{T} \alpha^k (c(i_k, \xi_{k-1}) + f(i_k, C_3 - \xi_{k-1})) \right) < C_4(T+1). \qquad (5.15)$$

To investigate the remaining term on the RHS of (5.14), it is convenient to introduce the notation

$$g(T, \alpha) := \mathsf{E}\left(\sum_{k=0}^{T} \alpha^k (f(i_k, C_3 - \sum_{t=0}^{k-1} \xi_t) \right).$$

We show that $g(\cdot, 1)$ is at least of quadratic growth. Let ε and ρ be the same as in Assumption (vii), and ν_h, ν_l, and N as in the proof of

Lemma 5.1. We split the sum with respect to k into sections of the length N and get

$$g(aN-1,1) := \sum_{k=0}^{a-1} \mathsf{E}\left(\sum_{j=kN}^{(k+1)N-1} (f(i_j, C_3 - \sum_{t=0}^{j-1} \xi_t)\right). \tag{5.16}$$

To obtain a lower bound for the argument of the first sum, we use the conditional expectation with respect to the event that the cumulative demand up to time kN is equal to z, i.e.,

$$\begin{aligned} &\mathsf{E}\left(\sum_{j=kN}^{(k+1)N-1} f(i_j, C_3 - \sum_{t=0}^{j-1} \xi_t)\right) \\ &= \int_0^\infty \mathsf{E}\left(\sum_{j=kN}^{(k+1)N-1} f(i_j, C_3 - \sum_{t=0}^{j-1} \xi_t) \,\middle|\, \sum_{t=0}^{kN-1} \xi_t = z\right) d\mathsf{P}\left(\sum_{t=0}^{kN-1} \xi_t < z\right). \end{aligned}$$

Because $j \geq kN$, we have $\sum_{t=0}^{j-1} \xi_t \geq \sum_{t=0}^{kN-1} \xi_t = z$. It follows from Assumption (ii) that $f(i_j, x)$ is decreasing for $x \leq 0$, and since $C_3 \leq 0$, we can conclude

$$\begin{aligned} &\mathsf{E}\left(\sum_{j=kN}^{(k+1)N-1} f(i_j, C_3 - \sum_{t=0}^{j-1} \xi_t)\right) \\ &\geq \int_0^\infty \mathsf{E}\left(\sum_{j=kN}^{(k+1)N-1} f(i_j, C_3 - z)\right) d\mathsf{P}\left(\sum_{t=0}^{kN-1} \xi_t < z\right). \end{aligned} \tag{5.17}$$

If we restrict our attention to those periods j in which $i_j = l$, we can replace $f(l, C_3 - z)$ by its lower bound qz, use (5.12), and obtain

$$\begin{aligned} &\mathsf{E}\left(\sum_{j=kN}^{(k+1)N-1} f(i_j, C_3 - z)\right) \\ &\geq \mathsf{E}\left(\sum_{j=kN}^{(k+1)N-1} f(i_j, C_3 - z)\mathbb{1}_{i_j=l}\right) \\ &= \mathsf{E}\left(\sum_{j=kN}^{(k+1)N-1} f(l, C_3 - z)\mathbb{1}_{i_j=l}\right) \\ &\geq qz \sum_{j=kN}^{(k+1)N-1} \mathsf{P}(i_j = l) \geq qz\nu_l/2. \end{aligned} \tag{5.18}$$

Substituting (5.18) into (5.17) yields

$$\mathsf{E}\left(\sum_{j=kN}^{(k+1)N-1} f(i_j, C_3 - \sum_{t=0}^{j-1} \xi_t)\right) \geq \frac{q\nu_l}{2}\mathsf{E}\left(\sum_{t=0}^{kN-1} \xi_t\right). \tag{5.19}$$

From (5.13) we can easily derive

$$\mathsf{E}\left(\sum_{t=aN}^{(a+1)N-1} \xi_t\right) = \mathsf{E}(\psi^a) \geq \nu_h \rho\varepsilon/2 > 0$$

and, therefore,

$$\mathsf{E}\left(\sum_{t=0}^{kN-1} \xi_t\right) \geq k\nu_h\rho\varepsilon/2. \tag{5.20}$$

Combining (5.16), (5.20), and (5.19) finally yields

$$g(aN-1, 1) \geq \sum_{k=0}^{a-1} k\nu_h\rho\varepsilon q\nu_l/4 = \nu_h\rho\varepsilon q\nu_l a(a-1)/8. \tag{5.21}$$

Because $g(\cdot, 1)$ is increasing, (5.21) establishes that it is at least of quadratic growth. But from this and (5.15), it follows that there is a T^* such that $g(T^*, 1) > C_4(T^* + 1)$. Because of the continuity of $g(T, \cdot)$, it follows that there is an $\alpha_0 < 1$ such that for all $\alpha \geq \alpha_0$, it holds that $g(T^*, \alpha) > C_4(T^* + 1)$. Therefore,

$$v^\alpha(i, C_3) - J^\alpha(i, C_3, \tilde{U}) > 0, \quad 1 > \alpha \geq \alpha_0,$$

for the policy $\tilde{U}$ defined by $T = T^*$. This leads to a contradiction that the policy U is optimal, and helps us to conclude that any strategy with $\max\{s_i^\alpha\} < C_2 = C_3 - T^*M$ cannot be optimal. □

LEMMA 5.3 *Under Assumptions* (i)-(iii), (v)-(viii), *and for* α_0 *obtained in Lemma* 5.2, $v^\alpha(i, \cdot)$ *is locally equi-Lipschitzian for* $\alpha \geq \alpha_0$, *i.e., for* $b > 0$ *there is a constant* $C_5 \geq 0$, *independent of* α, *such that*

$$|v^\alpha(i, x) - v^\alpha(i, \tilde{x})| \leq C_5|x - \tilde{x}| \quad \textit{for all} \quad x, \tilde{x} \in [-b, b], \quad \alpha \geq \alpha_0. \tag{5.22}$$

Proof. Consider the case $\tilde{x} \geq x$. Let us fix an $\alpha \geq \alpha_0$, use the optimal strategy U with initial surplus x, and the strategy $\tilde{U}$ defined by

$$\tilde{u}_k = [u_k - (\tilde{x}_k - x_k)]^+ = \begin{cases} 0 & \text{if} \quad u_k \leq \tilde{x}_k - x_k, \\ u_k + x_k - \tilde{x}_k & \text{if} \quad u_k > \tilde{x}_k - x_k, \end{cases}$$

with initial $\tilde{x}$. It is easy to see that if we define $\tilde{\tau} := \inf\{t : \tilde{u}_t > 0\}$, then

$$\tilde{u}_k = \begin{cases} 0 & \text{for} \quad 0 \le k \le \tilde{\tau} - 1, \\ u_k + x_k - \tilde{x}_k & \text{for} \quad k = \tilde{\tau}, \\ u_k & \text{for} \quad k \ge \tilde{\tau} + 1, \end{cases} \tag{5.23}$$

and, therefore,

$$\tilde{x}_{\tilde{\tau}+1} = \tilde{x}_{\tilde{\tau}} + \tilde{u}_{\tilde{\tau}} - \xi_{\tilde{\tau}} = x_{\tilde{\tau}} + u_{\tilde{\tau}} - \xi_{\tilde{\tau}} = x_{\tilde{\tau}+1}.$$

It follows that for $k \ge \tilde{\tau}$, the surplus levels $\tilde{x}_k$ and x_k are equal and the order quantities $\tilde{u}_k$ and u_k are equal.

Furthermore, it is easy to conclude that

$$0 \le \tilde{u}_k \le u_k \text{ and } |\tilde{x}_k - x_k| \le |\tilde{x} - x| \text{ for all } k.$$

From Assumptions (i) and (ii), we have

$$c(i_k, \tilde{u}_k) \le c(i_k, u_k) \text{ and } |f(i_k, \tilde{x}_k) - f(i_k, x_k)| \le C|\tilde{x} - x|.$$

Therefore,

$$\begin{aligned} & v^\alpha(i, \tilde{x}) - v^\alpha(i, x) \\ \le \;& J^\alpha(i, \tilde{x}; \tilde{U}) - J^\alpha(i, x; U) \\ = \;& \mathsf{E}\Big(\sum_{k=0}^{\tilde{\tau}} \alpha^k (f(i_k, \tilde{x}_k) - f(i_k, x_k) + c(i_k, \tilde{u}_k) - c(i_k, u_k))\Big) \\ \le \;& \mathsf{E}\Big(\sum_{k=0}^{\tilde{\tau}} \alpha^k C|\tilde{x} - x|\Big) \le \mathsf{E}(\tilde{\tau} + 1) C|\tilde{x} - x|. \end{aligned} \tag{5.24}$$

In order to prove that $\mathsf{E}(\tilde{\tau} + 1)$ is bounded, note from (5.23) that $\tilde{x}_k - x_k = \tilde{x} - x - \sum_{t=0}^{k-1} u_k$ for $k \le \tilde{\tau}$ and, therefore,

$$\tilde{\tau} = \inf\{k : u_k > \tilde{x}_k - x_k\} = \inf\{k : \sum_{t=0}^{k} u_k > \tilde{x} - x\}.$$

Now consider the inventory level $y_n = x_n + u_n$ after ordering in a period n, if there exists one, in which $i_n = i^*$ with $s^\alpha_{i^*} \ge C_2$. From the structure of the optimal strategy, we know that under the assumption $\alpha \ge \alpha_0$ we have $y_n \ge C_2$ (see Lemma 5.2), and therefore,

$$C_2 - 1 < y_n = x + \sum_{j=0}^{n} u_j - \sum_{j=0}^{n-1} \xi_j.$$

This implies that if $\sum_{j=0}^{n-1}\xi_j > \tilde{x} - C_2 + 1$ and $i_n = i^*$, then $\sum_{j=0}^{n} u_j > \tilde{x} - x$, and therefore $\tilde{\tau} \le n$, a.s. It is easy to see that from Lemma 5.1, Assumption (vi), and the fact that $x, \tilde{x} \in [-b, b]$, we have

$$\begin{aligned} \mathsf{E}(\tilde{\tau}) &\le \mathsf{E}[\inf\{k : \sum_{j=0}^{k-1}\xi_j \ge \tilde{x} - C_2 + 1 \text{ and } i_k = i^*\}] \\ &\le \mathsf{E}_i(\tau_{\tilde{x}-C_2+1} + \inf\{k \ge \tau_{\tilde{x}-C_2+1} : i_k = i^*\}) < \infty. \end{aligned}$$

This inequality, together with (5.24), implies that for $\alpha \ge \alpha_0$ and $\tilde{x} \ge x$, we have

$$v^\alpha(i,\tilde{x}) - v^\alpha(i,x) \le C_5|\tilde{x} - x|.$$

To complete the proof, it is sufficient to prove the above inequality for $\tilde{x} < x$. In this case, let us define

$$\tilde{u}_k = \begin{cases} u_k + x - \tilde{x} & \text{if } k = k^*, \\ u_k & \text{otherwise,} \end{cases}$$

where k^* is the first k with $u_k > 0$. It is obvious that

$$\begin{aligned} & v^\alpha(i,\tilde{x}) - v^\alpha(i,x) \\ &\quad\le J^\alpha(i,\tilde{x};\tilde{U}) - J^\alpha(i,x;U) \\ &\quad= \mathsf{E}\Big(\sum_{k=0}^{k^*}\alpha^k(f(i_k,\tilde{x}_k) - f(i_k,x_k)) + \alpha^{k^*}c_{i_{k^*}}(\tilde{u}_{k^*} - u_{k^*})\Big) \\ &\quad\le \mathsf{E}(k^*+1)C|\tilde{x} - x| + \max\{c_i : i \in \mathbb{I}\}|\tilde{x} - x| \\ &\quad\le C_5|\tilde{x} - x|, \end{aligned} \tag{5.25}$$

where the boundedness of $\mathsf{E}(k^*+1)$ can be proved in a similar manner as above. □

LEMMA 5.4 *Let Assumptions* (i)-(viii) *hold. For* α_0 *defined in Lemma* 5.2, *there are* $\alpha_1 \ge \alpha_0$ *and* $C_6 \ge 0$ *such that for* $\alpha \ge \alpha_1$ *and* $x \ge C_6$, *the function* $v^\alpha(i,x)$ *is increasing. Furthermore,* $v^\alpha(i,x)$ *is decreasing for* $x \le 0$.

Proof. We begin with the proof of the first part of the lemma. Because of Assumptions (ii) and (iv), there are constants $0 \le C_7 < \infty$ and $q > 0$ such that $f(g,x) \ge qx$ for all $x \ge C_7$. Let us fix a positive integer T and let $C_6 = C_7 + (T+1)M$. Recall from Assumption (viii) that M is the upper bound on the demand. Later we will specify a value of T, namely T^*, for which the assertion of the lemma holds.

Assume the initial surplus $x > C_6$ and apply the optimal policy $U = (u(i_0, x_0), u(i_1, x_1), \ldots)$, where $u(i, x)$ is an optimal stationary feedback control and x_k is the associated surplus in the beginning of period k. Since $x > C_6$, there is an $\tilde{x}$ such that $x > \tilde{x} = x - \varepsilon > C_6$ and $1 \geq \varepsilon > 0$. Let us apply the policy $\tilde{U}$, with the initial surplus $\tilde{x}$, defined by

$$\tilde{u}_k = u(i_k, x_k) \text{ for } k < \tau \text{ and } \tilde{u}_k = u(i_k, \tilde{x}_k) \text{ for } k \geq \tau,$$

where $\tau := \inf\{k : \tilde{x}_k < C_7 + M\}$, with $\tilde{x}_k$ denoting the surplus in period k corresponding to the initial $\tilde{x}$ and the policy $\tilde{U}$. It is clear that $\tilde{U}$ orders the same amount in each of the periods from 0 to $\tau - 1$ as those ordered under the optimal strategy with the initial value $x_0 = x$. Beginning with τ, $\tilde{U}$ follows the optimal feedback policy.

For $k = 0, 1, \ldots, \tau$, the difference of the inventory processes $x_k - \tilde{x}_k$ remains ε and $\tilde{x}_k \geq C_7$ in view of the fact that $\xi_{k-1} < M$. Furthermore, the bound M on periodic demands implies $\tau > T$. Thus,

$$\begin{aligned} & v^\alpha(i, x) - v^\alpha(i, \tilde{x}) \\ \geq \; & v^\alpha(i, x) - J^\alpha(i, x - \varepsilon; \tilde{U}) \\ = \; & \sum_{k=0}^{\infty} \alpha^k \mathsf{E}(f(i_k, x_k) - f(i_k, \tilde{x}_k) + c(i_k, u(i_k, x_k) - c(i_k, \tilde{u}_k)) \\ = \; & \sum_{k=0}^{\tau-1} \alpha^k \mathsf{E}(f(i_k, x_k) - f(i_k, \tilde{x}_k)) + \mathsf{E}(\alpha^\tau (v^\alpha(i_\tau, x_\tau) - v^\alpha(i_\tau, \tilde{x}_\tau))). \end{aligned}$$

Since $v^\alpha(i, \cdot)$ is locally equi-Lipschitzian, $x_\tau - \tilde{x}_\tau = \varepsilon$, and

$$C_7 + M + 1 \geq C_7 + M + \varepsilon > x_\tau > \tilde{x}_\tau \geq C_7,$$

there is a nonnegative constant C_8 independent of x and ε such that

$$\mathsf{E}(\alpha^\tau (v^\alpha(i_\tau, x_\tau) - v^\alpha(i_\tau, \tilde{x}_\tau))) \geq -C_8 \varepsilon \quad \text{for all } \alpha_0 \leq \alpha < 1.$$

For further analysis, set

$$h(t, \alpha) := \sum_{k=0}^{t-1} \alpha^k \mathsf{E}(f(i_k, x_k) - f(i_k, \tilde{x}_k)), \quad t \geq 1.$$

Furthermore, because $\tau \geq T + 1$ and $x_k > \tilde{x}_k$ for $k \leq \tau$, we have for the state g specified in Assumption (iv),

$$\begin{aligned} h(\tau, 1) \; & \geq \; h(T+1, 1) \geq \sum_{k=0}^{T} \mathsf{E}((f(i_k, x_k) - f(i_k, \tilde{x}_k)) \mathbb{1}_{i_k = g}) \\ & = \; \sum_{k=0}^{T} \mathsf{E}((f(g, x_k) - f(g, \tilde{x}_k)) \mathbb{1}_{i_k = g}) \geq q\varepsilon \sum_{k=0}^{T} \mathsf{P}(i_k = g). \end{aligned}$$

On the other hand, with the notation of the proof of Lemma 5.1, we know that

$$aN \geq \sum_{k=0}^{aN-1} \mathsf{P}(i_k = g) = \sum_{b=0}^{a-1} \sum_{k=bN}^{(b+1)N-1} \mathsf{P}(i_k = g) \geq a\nu_g/2$$

and, therefore, $q\varepsilon \sum_{k=0}^{T} \mathsf{P}(i_k = g)$ is of linear growth in T. But then, there is a T^* such that

$$q\varepsilon \sum_{k=0}^{T^*} \mathsf{P}(i_k = g) > C_8\varepsilon,$$

which implies the strict inequality $h(T^* + 1, 1) > C_8\varepsilon$. Because $h(t, \cdot)$ is continuous, we know that there is an $\alpha_1 \geq \alpha_0$ such that the strict inequality holds for all $1 \geq \alpha \geq \alpha_1$, and we have

$$v^\alpha(i, x) - v^\alpha(i, x - \varepsilon) \geq 0 \quad \text{for } x \geq C_6 \text{ and } 1 \geq \alpha \geq \alpha_1.$$

The second part of the lemma claiming that $v^\alpha(i, x)$ is decreasing for $x \leq 0$ is easy to prove. We only want to sketch the proof here. Let $x < \tilde{x} \leq 0$ and let U denote an optimal decision used with the initial value x. Let $\tilde{U}$ be a strategy applied to the initial value $\tilde{x}$ according to

$$\tilde{u}_k := [u_k - (\tilde{x}_k - x_k)]^+ \text{ for } k = 0, 1, \ldots.$$

Then, it is easy to see that $\tilde{u}_k \leq u_k$, $\tilde{x}_k \geq x_k$ if $\tilde{x}_k \leq \tilde{x} \leq 0$, and $\tilde{x}_k = x_k$ if $\tilde{x}_k > \tilde{x}$.

Thus, the control cost is not higher, the shortage is not larger, and the inventory is equal under $\tilde{u}$ when compared to u. Therefore, $v^\alpha(i, \tilde{x}) \leq J^\alpha(i, \tilde{x}; \tilde{U}) \leq v^\alpha(i, x)$, which is the required monotonicity property. □

5.5. Vanishing Discount Approach

LEMMA 5.5 *Under Assumptions* (i)-(iii), (v)-(viii), *and for* α_0 *obtained in Lemma* 5.2, *the differential discounted value function* $w^\alpha(i, x) := v^\alpha(i, x) - v^\alpha(1, 0)$ *is uniformly bounded with respect to* $\alpha \geq \alpha_0$ *for all* x *and* i.

Proof. Since Lemma 5.3 implies

$$\begin{aligned} |w^\alpha(i, x)| &= |v^\alpha(i, x) - v^\alpha(1, 0)| \\ &\leq |v^\alpha(i, x) - v^\alpha(i, 0)| + |v^\alpha(i, 0) - v^\alpha(1, 0)| \\ &\leq C_5|x| + |w^\alpha(i, 0)|, \end{aligned}$$

it is sufficient to prove that $w^\alpha(i,0)$ is uniformly bounded. Note that C_5 may depend on x, but it is independent of $\alpha \geq \alpha_0$.

First, we show that there is an $\underline{M} > -\infty$ with $w^\alpha(i,0) \geq \underline{M}$ for all $\alpha \geq \alpha_0$. Let α be fixed. From Theorem 5.1, we know that in this discounted case there is a stationary optimal feedback policy $U = (u(i,x), u(i,x), \ldots)$. With $k^* = \inf\{k : i_k = i\}$, we consider the cost for the initial state $(i_0, \tilde{x}_0) = (1, 0)$, and the following inventory policy $\tilde{U}$:

$$\tilde{u}_k = \xi_{k-1} \text{ for } k < k^*,\ \tilde{u}_{k^*} = u(i_{k^*}, 0) - x_{k^*} = u(i_{k^*}, 0) + \xi_{k^*-1},$$
$$\text{and } \tilde{u}_k = u(i_k, \tilde{x}_k) \text{ for } k > k^*.$$

That is, we order the previous period demand until the demand state is equal to i in period k^*. In that period, we order as much as is needed to reach an inventory level after ordering, that is equal to the inventory level in period 0 after ordering the optimal amount $u(i,0)$ for the problem starting in the state $(i,0)$. From that instant on, we apply the optimal feedback policy U.

The cost corresponding to this policy is

$$J^\alpha(1,0;\tilde{U}) = \mathsf{E}\left(\sum_{k=0}^{k^*-1} \alpha^k (c(i_k, \tilde{u}_k) + f(i_k, \tilde{x}_k)) + \alpha^{k^*} v^\alpha(i, -\xi_{k^*-1})\right). \quad (5.26)$$

To derive an upper bound for $\mathsf{E}(v^\alpha(i, -\xi_{k^*-1}))$ in terms of $v^\alpha(i,0)$, we use the dynamic programming equation (5.9). For $y \geq 0$, we have

$$v^\alpha(i,-y) \leq f(i,-y) + c(i, u(i,0)+y) + \alpha \sum_{j=1}^{L} p_{ij} \int_0^M v^\alpha(j, u(i,0)-z)\varphi_i(z)dz.$$

On the other hand, because U is optimal, we have

$$v^\alpha(i,0) = f(i,0) + c(i, u(i,0)) + \alpha \sum_{j=1}^{L} p_{ij} \int_0^M v^\alpha(j, u(i,0) - z)\varphi_i(z)dz.$$

Thus,

$$\begin{aligned} v^\alpha(i,-y) \ \leq\ & v^\alpha(i,0) + f(i,-y) + c(i, u(i,0) + y) \\ & -f(i,0) - c(i, u(i,0)). \end{aligned} \quad (5.27)$$

We use (5.27) with $y = \xi_{k^*-1}$, use $\tilde{u}_k = \xi_{k-1}$ and $\tilde{x}_k = -\xi_{k-1}$ for $k < k^*$ in (5.26), and combine them to obtain

$$J^\alpha(1,0;\tilde{U}) \ \leq\ \mathsf{E}\Big(\sum_{k=0}^{k^*-1} \alpha^k (c(i_k, \xi_{k-1}) + f(i_k, -\xi_{k-1})) + \alpha^{k^*} v^\alpha(i,0)$$

$$+\alpha^{k^*}(c(i,u(i,0)+\xi_{k^*-1})+f(i,-\xi_{k^*-1})$$
$$-c(i,u(i,0))-f(i,0))\Big). \tag{5.28}$$

Because of the bounded demand, all arguments of the sums in the first line and the expressions on the second line are uniformly bounded from above by a constant $\underline{M}$. Moreover, $\mathsf{E}(k^*)<\infty$ because of Assumption (vi). Therefore, it holds that

$$\begin{aligned} w^\alpha(i,0) &= v^\alpha(i,0)-v^\alpha(1,0)\geq v^\alpha(i,0)-J^\alpha(1,0;\tilde{U}) \\ &\geq v^\alpha(i,0)-\mathsf{E}\Big(\sum_{k=0}^{k^*-1}\alpha^k(c(i_k,\xi_{k-1})+f(i_k,-\xi_{k-1})) \\ &\quad +\alpha^{k^*}v^\alpha(i,0)+\alpha^{k^*}(c(i,u(i,0)+\xi_{k^*-1}) \\ &\quad +f(i,-\xi_{k^*-1})-c(i,u(i,0))-f(i,0))\Big) \\ &\geq v^\alpha(i,0)(1-\mathsf{E}(\alpha^{k^*}))+\underline{M}\geq\underline{M}. \end{aligned} \tag{5.29}$$

The opposite inequality, $w^\alpha(i,0)\leq\overline{M}$, is shown analogously by changing the role of the states 1 and i. Thus, $|w^\alpha(i,x)|\leq C_5|x|+\max\{\underline{M},\overline{M}\}$, and the proof is completed. □

LEMMA 5.6 *Under Assumption* (viii), $(1-\alpha)v^\alpha(1,0)$ *is uniformly bounded on* $0<\alpha<1$.

Proof. Consider the strategy $\tilde{U}=(0,\xi_0,\xi_1,\ldots)$. Then, because $\tilde{U}$ is not necessarily optimal,

$$0\leq v^\alpha(1,0)\leq J^\alpha(1,0;\tilde{U})=\mathsf{E}\left(\sum_{k=0}^{\infty}\alpha^k(c(i_k,\xi_{k-1})+f(i_k,-\xi_{k-1}))\right).$$

Because the demand is bounded, there is a $C_9<\infty$ such that

$$\mathsf{E}(c(i_k,\xi_{k-1})+f(i_k,-\xi_{k-1}))<C_9.$$

Therefore,

$$\begin{aligned} 0\leq(1-\alpha)v^\alpha(1,0) &\leq (1-\alpha)\sum_{k=0}^{\infty}\alpha^k\mathsf{E}(c(i_k,\xi_{k-1})+f(i_k,-\xi_{k-1})) \\ &\leq (1-\alpha)\sum_{k=0}^{\infty}\alpha^kC_9=C_9. \end{aligned}$$

□

THEOREM 5.2 *Let Assumptions* (i)-(iii) *and* (v)-(viii) *hold. There exist a sequence* $(\alpha_k)_{k=1}^{\infty}$ *converging to* 1, *a constant* λ^*, *and a locally Lipschitz continuous function* $w^*(\cdot,\cdot)$ *such that*

$$(1-\alpha_k)v^{\alpha_k}(i,x) \to \lambda^* \quad \textit{and} \quad w^{\alpha_k}(i,x) \to w^*(i,x),$$

locally uniformly in x *and* i *as* k *goes to infinity. Moreover,* (λ^*, w^*) *satisfies the average cost optimality equation*

$$w(i,x) + \lambda = f(i,x) + \inf_{u\geq 0}\{c(i,u) + F(w)(i,x+u)\}. \tag{5.30}$$

Proof. It is immediate from Lemma 5.3 and the definition of $w^\alpha(i,x)$ that $w^\alpha(i,\cdot)$ is locally equi-Lipschitzian for $\alpha \geq \alpha_0$, and therefore it is uniformly equicontinuous in any finite interval by Theorem A.3.2. Additionally, according to Lemma 5.5, $w^\alpha(i,\cdot)$ is uniformly bounded, and by Lemma 5.6, $(1-\alpha)v^\alpha(1,0)$ is uniformly bounded. Therefore, from the Arzelà-Ascoli Theorem A.3.5 and Lemma 5.3, there is a sequence $\alpha_k \to 1$, a locally Lipschitz continuous function $w^*(i,x)$, and a constant λ^* such that

$$(1-\alpha_k)v^{\alpha_k}(1,0) \to \lambda^*, \quad \text{and} \quad w^{\alpha_k}(i,x) \to w^*(i,x)$$

for each x locally uniformly in any given interval. By the diagonalization procedure, a subsequence can be found so that $w^{\alpha_k}(i,\cdot)$ converges to a locally Lipschitz continuous function $w^*(i,\cdot)$ on the entire real line; (see Kirszbraun Theorem A.1.4).

Next, it is easy to see that

$$\lim_{k\to\infty}(1-\alpha_k)v^{\alpha_k}(i,x) = \lim_{k\to\infty}(1-\alpha_k)(w^{\alpha_k}(i,x) + v^{\alpha_k}(1,0)) = \lambda^*.$$

Substituting $v^{\alpha_k}(i,x) = w^{\alpha_k}(i,x) + v^{\alpha_k}(1,0)$ in (5.9) yields

$$\begin{aligned} &w^{\alpha_k}(i,x) + (1-\alpha_k)v^{\alpha_k}(1,0) \\ &\quad = f(i,x) + \inf_{u\geq 0}\{c(i,u) + F(w^{\alpha_k})(i,x+u)\}. \end{aligned} \tag{5.31}$$

Since $w^{\alpha_k}(i,x)$ converges locally uniformly with respect to x and i and since for a given x, $x+u-\xi \in [x-M, \max\{x, C_6+M\}]$ by Lemma 5.4, we can pass to the limit on both sides of (5.31) and obtain (5.30); (see Theorem A.2.4). This completes the proof. □

5.6. Verification Theorem

DEFINITION 5.1 Let (λ, w) be a solution of the average optimality equation (5.30). An admissible strategy $U = (u_0, u_1, \ldots)$ is called *stable* with respect to w if for each initial surplus level x and for each $i \in \mathbb{I}$,

$$\lim_{k\to\infty} \frac{1}{k}\mathsf{E}(w(i_k, x_k)) = 0,$$

where x_k is the surplus level in period k corresponding to the initial state (i, x) and the strategy U. Here α is called the average cost and w is called a potential function.

THEOREM 5.3 (Verification Theorem)
(a) *Let $(\lambda, w(\cdot,\cdot))$ be a solution of the average cost optimality equation* (5.30). *Then, $\lambda \le J(i, x; U)$ for any admissible U.*
(b) *If it exists, let $u^*(i, x)$ attain the infimum in* (5.30). *Furthermore, let $U^* = (u^*, u^*, \ldots)$, the stationary feedback policy given by u^*, be stable with respect to w. Then,*

$$\lambda = J(i, x; U^*) = \lambda^* = \lim_{N\to\infty} \frac{1}{N}\mathsf{E}\left(\sum_{k=0}^{N-1} f(i_k, x_k^*) + c(i_k, u_k^*)\right),$$

and U^* is an average optimal strategy.
(c) *Moreover, U^* minimizes*

$$\liminf_{N\to\infty} \frac{1}{N}\mathsf{E}\left(\sum_{k=0}^{N-1} f(i_k, x_k) + c(i_k, u_k)\right)$$

over the class of admissible decisions which are stable with respect to w.

Proof. Let $U = (u_0, u_1, \ldots)$ denote any admissible decision. Suppose

$$J(i, x; U) < \lambda. \tag{5.32}$$

Put $\tilde{f}(k) = \mathsf{E}[f(i_k, x_k) + c(i_k, u_k)]$. From (5.32), it immediately follows that $\sum_{k=0}^{n-1} \tilde{f}(k) < \infty$ for each positive integer n, since otherwise we would have $J(i, x; U) = \infty$. Note that

$$J(i, x; U) = \limsup_{n\to\infty} \frac{1}{n}\sum_{k=0}^{n-1} \tilde{f}(k),$$

while

$$(1-\alpha)J^{\alpha}(i, x; U) = (1-\alpha)\sum_{k=0}^{\infty} \alpha^k \tilde{f}(k). \tag{5.33}$$

Since $\tilde{f}(k)$ is nonnegative for each k, the sum in (5.33) is well defined for $0 \leq \alpha < 1$, and we can use the Tauberian Theorem A.5.2 to obtain

$$\limsup_{\alpha \uparrow 1} (1-\alpha) J^{\alpha}(i,x;U) \leq J(i,x;U) < \lambda.$$

On the other hand, we know from Theorem 5.2 that $(1-\alpha_k)v^{\alpha_k}(i,x) \to \lambda$ on a subsequence $\{\alpha_k\}_{k=1}^{\infty}$ converging to one. Thus, there exists an $\alpha < 1$ such that

$$(1-\alpha) J^{\alpha}(i,x;U) < (1-\alpha) v^{\alpha}(i,x),$$

which contradicts the definition of the value function $v^{\alpha}(i,x)$. This proves part (a) of the theorem.

To prove part (c), we assume that U is stable with respect to w, and then follow the same approach used in deriving (2.15) to obtain

$$\mathsf{E}\{w(i_{k+1}, x_{k+1}) \,|\, i_0, \ldots, i_k, \xi_0, \ldots, \xi_{k-1}\} = F(w)(i_k, x_k + u_k) \quad \text{a.s.} \tag{5.34}$$

Because u_k does not necessarily attain the infimum in (5.30), we have

$$w(i_k, x_k) + \lambda \leq f(i_k, x_k) + c(i_k, u_k) + F(w)(i_k, x_k + u_k) \quad \text{a.s.},$$

and from (5.34) we derive

$$w(i_k, x_k) + \lambda \leq f(i_k, x_k) + c(i_k, u_k) + \mathsf{E}(w(i_{k+1}, x_k + u_k - \xi_k) \,|\, i_k) \quad \text{a.s.}$$

By taking the expectation on both sides, we obtain

$$\mathsf{E}(w(i_k, x_k)) + \lambda \leq \mathsf{E}(f(i_k, x_k) + c(i_k, u_k)) + \mathsf{E}(w(i_{k+1}, x_{k+1})).$$

Summing from 0 to $n-1$ yields

$$n\lambda \leq \mathsf{E}\left(\sum_{k=0}^{n-1} f(i_k, x_k) + c(i_k, u_k) \right) + \mathsf{E}(w(i_n, x_n)) - \mathsf{E}(w(i_0, x_0)). \tag{5.35}$$

Divide by n, let n go to infinity, and use the fact that U is stable with respect to w, to obtain

$$\lambda \leq \liminf_{n \to \infty} \frac{1}{n} \mathsf{E}\left(\sum_{k=0}^{n-1} f(i_k, x_k) + c(i_k, u_k) \right). \tag{5.36}$$

On the other hand, if there exists a u^* which attains the infimum in (5.30), we then have

$$w(i_k, x_k) + \lambda = f(i_k, x_k) + c(i_k, u^*(i_k, x_k)) + F(w)(i_k, x_k + u^*(i_k, x_k)), \text{ a.s.},$$

and we analogously obtain

$$n\lambda = \mathsf{E}\left(\sum_{k=0}^{n-1} f(i_k, x_k) + c(i_k, u^*(i_k, x_k))\right) + \mathsf{E}(w(i_n, x_n)) - \mathsf{E}(w(i_0, x_0)). \tag{5.37}$$

Because U^* is assumed stable with respect to w, we get

$$\lambda = \lim_{n\to\infty} \frac{1}{n}\mathsf{E}\left(\sum_{k=0}^{n-1} f(i_k, x_k) + c(i_k, u^*(i_k, x_k)\right) = J(i, x; U^*). \tag{5.38}$$

This proves part (b) of the theorem. □

REMARK 5.4 It should be obvious that any solution (λ, w) of the average cost optimality equation and control u^* satisfying (b) of Theorem 5.3 will have a unique λ, since it represents the minimum average cost. On the other hand, if (λ, w) is a solution, then $(\lambda, w + c)$, where c is any constant, is also a solution. For the purpose of this chapter, we do not require w to be unique up to a constant. If w is not unique up to a constant, then u^* may not be unique. We also do not need w^* in Theorem 5.2 to be unique.

LEMMA 5.7 *The function $w^*(i, \cdot)$, defined in Theorem 5.2, is K-convex.*

Proof. We know from Chapter 2 that $v^\alpha(i, \cdot)$ is K-convex for each $\alpha < 1$. Therefore, the function $w^{\alpha_k}(i, \cdot)$ is K-convex for each k. It follows that $w^*(i, \cdot)$, the limit of a sequence of K-convex functions, is itself K-convex. □

THEOREM 5.4 *Let Assumptions* (i)-(viii) *hold. There are constants $\infty \geq S_i \geq 0$ and $s_i \leq S_i$, $i \in \mathbb{I}$ satisfying $\max_{i\in\mathbb{I}}\{s_i\} \geq C_2$, with C_2 specified in Lemma 5.2, such that the stationary feedback strategy $U^* = (u^*, u^*, \ldots)$, according to $u^*(i, x) = (S_i - x)\mathbb{1}_{x<s_i}$, is average optimal.*

Proof. The proof is based on verifying that (λ^*, w^*) obtained in Theorem 5.2 satisfies the conditions of Theorem 5.3 and that $w^*(i, \cdot)$ is K-convex according to Lemma 5.7. The details follow.

It immediately follows from Lemma 5.4 that $w^\alpha(i, x)$ is decreasing for $x \leq 0$ and increasing for $x \geq C_6$ for $\alpha_1 \leq \alpha < 1$. This implies the same property for $w^*(i, x)$. Furthermore, by Theorem 5.2, $w^*(i, \cdot)$ is locally Lipschitz continuous. Therefore, $w^*(i, \cdot)$ is bounded from below.

Also, by Theorem 5.2, (λ^*, w^*) satisfies the average cost optimality equation (5.30), and $w^*(i, \cdot)$ is of linear growth.

Because $w^*(i, \cdot)$ is K-convex, we know that a minimizer in (5.30) is given by $u^*(i, x) = (S_i - x)\mathbb{1}_{x<s_i}$, where S_i minimizes $G(i, y) := c_i y +$

$F(w)(i,y)$, and s_i is a solution of the equation $G(i,y) = K + G(i,S_i)$ in y, if one exists, or $s_i = -\infty$ otherwise.

To prove that the strategy U^* is stable with respect to w, we derive bounds for S_i and s_i. Obviously, S_i as a global minimizer of $c_i y + F(w^*)(i,y)$ can be chosen to be less than or equal to $C_6 + M$, because $w^*(i,x)$ is increasing for $x \geq C_6$, which implies that $F(w^*)(i,y)$ is increasing for $y \geq C_6 + M$.

For $\alpha_0 \leq \alpha < 1$, an (s_i^α, S_i^α) strategy is optimal, $v^\alpha(i,\cdot)$ is K-convex, and, from Lemma 5.2 there is a $j \in \mathbb{I}$ such that $s_j^\alpha \geq C_2$. We denote the smallest such j by $j(\alpha)$. For the sequence $(\alpha_k)_{k=1}^\infty$ from Theorem 5.2, at least one j^* appears infinitely often among the $j(\alpha_k)$. In the following, we choose a subsequence, still denoted by $(\alpha_k)_{k=1}^\infty$, such that $j(\alpha_k) = j^*$. Then, for all $1 > \alpha_k \geq \alpha_0$, we know that $s_{j^*}^\alpha \geq C_2$ and $v^{\alpha_k}(j^*,x)$ are decreasing for $x \leq s_{j^*}$ on account of the K-convexity of $v^{\alpha_k}(j^*,x)$ given in Theorem 5.1. Furthermore, $v^\alpha(j^*,x)$ is increasing for $x \geq C_6$ for $\alpha_1 \leq \alpha < 1$, $F(v^{\alpha_k})(j^*,y)$ is increasing for $y \geq C_6 + M$, and the infimum is attained for $y \leq C_6 + M$. By the definition of s_i^α, we know that the minimum of $v^\alpha(i,x)$ is taken at a value not less than s_i^α, and therefore at or above C_2. Thus, we can conclude that for $1 > \alpha_k > \max\{\alpha_0, \alpha_1\}$, we have

$$c_{j^*}C_2 + \alpha_k F(v^{\alpha_k})(j^*,C_2) - \min_{C_2 \leq y \leq C_6+M}\{c_{j^*}y + \alpha_k F(v^{\alpha_k})(j^*,y)\} \geq K.$$

Subtracting $\alpha_k v^{\alpha_k}(1,0)$ from both the terms on the LHS yields

$$c_{j^*}C_2 + \alpha_k F(w^{\alpha_k})(j^*,C_2) - \min_{C_2 \leq y \leq C_6+M}\{c_{j^*}y + \alpha_k F(w^{\alpha_k})(j^*,y)\} \geq K.$$

The functions $w^{\alpha_k}(j^*,y)$ converge uniformly in any closed interval to $w^*(j^*,y)$. Therefore, we can pass to the limit and get

$$c_{j^*}C_2 + F(w^*)(j^*,C_2) - \min_{C_2 \leq y \leq C_6+M}\{c_{j^*}y + F(w^*)(j^*,y)\} \geq K. \quad (5.39)$$

From (5.39) and the continuity of $w^*(j^*,\cdot)$, it follows that we can choose an $s_{j^*} \in [C_2, S_{j^*}]$ such that

$$c_{j^*}s_{j^*} + F(w^*)(j^*,s_{j^*}) - c_{j^*}S_{j^*} - F(w^*)(j^*,S_{j^*}) = K.$$

This means that under U^*, there will be some ordering infinitely often. Because of the structure of the strategy, the inventory level under U^* will never be higher than $\max\{x, S_1, \ldots, S_L\}$. Therefore, we have

$$x_k \leq \max\{x, C_6 + M\}. \quad (5.40)$$

Furthermore, the inventory level is at its lowest between any two consecutive orders right before the second order. Similar to the proof of

Lemma 5.1, it is easy to see that $\mathsf{E}(\tau_0)$ is finite, where τ_0 is the time between any two consecutive orders.

On the other hand, the expected demand between any two consecutive orders is less than or equal to $M(1+\mathsf{E}(\tau_0))$. It follows that

$$\lim_{k\to\infty} \mathsf{E}(x_k) \geq \min\{S_1, S_2, \ldots, S_L\} - M(1+\mathsf{E}(\tau_0)). \tag{5.41}$$

Furthermore, setting $u = -x$ in (5.30) we get

$$w^*(i,x) + \lambda \leq f(i,x) + c(i,-x) + F(w^*)(i,0), \quad x < 0.$$

Therefore, $w^*(i,x)$ is of at most linear growth for $x \to -\infty$. Because $w^*(i,x)$ is decreasing as x increases for $x \leq 0$ by Lemma 5.6, we can finally conclude from (5.40) and (5.41) that U^* is stable with respect to w^*.

We have now shown that the policy U^* satisfies all the conditions of the verification theorem (Theorem 5.3), and therefore it minimizes the long-run average cost. □

5.7. Concluding Remarks and Notes

In this chapter based on Beyer and Sethi (1997), we have carried out a rigorous analysis of the average cost stochastic inventory problem with Markovian demand and fixed ordering cost. We have proved a verification theorem for the average cost optimality equation, which we have used to establish the existence of an optimal state-dependent (s, S) policy.

The chapter generalizes the earlier studies of Iglehart (1963b), Veinott and Wagner (1965), Zheng (1991), and others in several ways. We use Markovian demand and convex surplus cost instead of independent demand and linear surplus cost, as assumed in the classical literature. Our Assumption (v) generalizes the usual assumption $p > c$ in the classical literature; (see Remark 2.1 in Chapter 2 for details). The vanishing discount approach used here does not require the stationary analysis carried out in Iglehart (1963b), which is not possible in all but the simplest of the cases. Instead, the approach requires solving the average cost optimality equation for obtaining the optimal solution.

We have, however, made the assumption that the demand is bounded and the costs are of linear growth. One reason for making the assumption was to simplify the mostly constructive proofs obtained in this chapter. Nevertheless, we relax these assumptions in the next chapter.

The vanishing discount approach has been also used very effectively to study continuous-time stochastic manufacturing systems. The reader is referred to Sethi *et al.* (2005b).

Chapter 6

AVERAGE COST MODELS WITH POLYNOMIALLY GROWING SURPLUS COST

With the results in hand for the discounted cost models with polynomially growing surplus cost as described in Chapter 3, we can now begin the analysis of the corresponding average cost model.

This chapter extends the model of Chapter 5 by allowing for unbounded demand and polynomially growing cost. To accomplish these relaxations, we impose a condition on a particular moment of the demand to be finite. The plan of the chapter is a follows. In Section 6.1, we provide a precise formulation of the problem, while relying on the notations introduced in Chapters 3 and 5. In Section 6.2, we derive relevant properties of the discounted cost value function that are uniform in the discount factor α, and examine the asymptotic behavior of the differential discounted value function as α approaches one. In Section 6.3, we employ the vanishing discount approach to establish the average cost optimality equation. The associated verification theorem is proved in Section 6.4, and the theorem is used to show that there exists an optimal policy characterized by parameters (s_i, S_i), $i \in \mathbb{I}$, where $\mathbb{I}$ denotes a finite collection of possible demand states. Moreover, these parameters can be obtained by taking the limit of a converging subsequence of (s_i^α, S_i^α) as $\alpha \to 1$.

6.1. Formulation of the Problem

In what follows, we show the existence of an optimal (s_i, S_i) policy for the stationary long-run average cost problem. We use the notation introduced at the beginning of Section 5.2. To establish this result we have to make assumptions in addition to those already made in Section 3.2 of Chapter 3. For the convenience of the reader, we list all the required assumptions here.

D. Beyer et al., *Markovian Demand Inventory Models*, International Series in Operations Research & Management Science 108,
DOI 10.1007/978-0-387-71604-6_6,

(i) The production cost is given by $c(i, u) = K + c_i u$ for $u > 0$ and $c(i, 0) = 0$, where the fixed ordering cost $K \geq 0$ and the variable cost $c_i \geq 0$.

(ii) For each i, the nonnegative surplus cost function $f(i, \cdot)$ is convex and at most of polynomial growth with power γ, i.e., $f(i, x) \leq \bar{f}(1 + |x|^\gamma)$ for some $\bar{f} > 0$.

(iii) There is a state $l \in \mathbb{I}$ such that $f(l, x)$ is not identically zero for $x \leq 0$.

(iv) There is a state $g \in \mathbb{I}$ such that $f(g, x)$ is not identically zero for $x \geq 0$.

(v) The production and inventory costs satisfy

$$c_i x + \sum_{j=1}^{L} p_{ij} \int_0^\infty f(j, x - z) d\Phi_i(z) \to \infty \text{ as } x \to \infty.$$

(vi) The Markov chain $(i_k)_{k=0}^\infty$ is irreducible.

(vii) There is a state $h \in \mathbb{I}$ such that $1 - \Phi_h(\varepsilon) = \rho > 0$ for some $\varepsilon > 0$.

(viii) There exists an $\varepsilon > 0$ such that $\mathsf{E}(\xi_k)^{\gamma+\varepsilon} \leq D < \infty$ for all k.

REMARK 6.1 Assumptions (i), (ii), and (v) are the same as the ones needed in Sections 3.5 and 3.6 of Chapter 3 to establish the optimality of a stationary (s_i^α, S_i^α) strategy in the stationary discounted infinite-horizon problem; (see Remarks 3.1 and 3.2).

REMARK 6.2 Assumptions (iii) and (iv) rule out trivial cases which lead to degenerate optimal policies. In fact, if Assumption (iii) is violated, the optimal policy is never to order. If Assumption (iii) holds and Assumption (iv) is violated, then it is optimal to wait for a period with a demand state for which c_i is minimal, and order an infinite amount in that period. Assumptions (iv) and (v) generalize the usual assumption made by Scarf (1960) and others that the unit inventory holding cost $h > 0$.

REMARK 6.3 Assumptions (vi)-(viii) are needed in order to guarantee that the exogenous demand depletes any given initial inventory in a finite expected time. While Assumption (vii) says that in at least one state h, the expected demand is strictly larger than zero, Assumption (vi) implies that the state h would occur infinitely often with a finite expected interval between successive occurrences. Assumption (viii) is technical.

It ensures finiteness of various expectations involved in the analysis of problems with the long-run average cost criterion.

The objective is to minimize the expected long-run average cost

$$J(i,x;U) = \limsup_{N\to\infty} \frac{1}{N}\mathsf{E}\left\{\sum_{k=0}^{N-1}[c(i_k,u_k)+f(i_k,x_k)]\right\}, \tag{6.1}$$

with $i_0 = i$ and $x_0 = x$, where as before $U = (u_0, u_1, \ldots)$, $u_i \geq 0$, $i = 0, 1, \ldots$, is a *history-dependent* or *nonanticipative* decision (order quantities) for the problem. Such a control U is termed admissible. Let $\mathcal{U}$ denote the class of all admissible controls. The surplus balance equations are given by

$$x_{k+1} = x_k + u_k - \xi_k, \; k = 0, 1, \ldots.$$

Our aim is to show that there exist a constant λ^*, termed the optimal average cost, which is independent of the initial i and x, and a control $U^* \in \mathcal{U}$ such that

$$\lambda^* = J(i,x;U^*) \leq J(i,x;U), \quad \forall \quad U \in \mathcal{U},$$

and

$$\lambda^* = \lim_{N\to\infty} \frac{1}{N}\mathsf{E}\left\{\sum_{k=0}^{N-1}[c(i_k,u_k^*)+f(i_k,x_k^*)]\right\},$$

where x_k^*, $k = 0, 1, \ldots$, is the surplus process corresponding to U^* with $i_0 = i$ and $x_0 = x$.

To prove these results, we will use the vanishing discount approach. That is, by letting the discount factor α in the discounted cost problem approach one, we will show that we can derive a dynamic programming equation whose solution provides an average optimal control and the associated minimum average cost λ^*.

6.2. Behavior of the Discounted Cost Model with Respect to the Discount Factor

As a first step in the vanishing discount approach, we will establish a uniform lower bound for the largest of the optimal ordering levels s_i^α with respect to α. We will do so by comparing the cost of two policies. The first one is an (s_i^α, S_i^α) policy with s_i^α less than a negative real number $-y$. In the proof of Lemma 6.3, we will give a modified policy and prove that there is a value y^* for y that is independent of the values of α close to one, such that this modified policy produces lower cost than the original policy. This will prove that any policy with $s_i^\alpha < -y^*$ for all i and α in the neighborhood of one cannot be optimal. To do so we need to obtain some preliminary results.

For convenience of exposition, let us define $x_0^0 = 0$ and x_k^0 as the surplus level at the beginning of period k corresponding to the policy of not ever ordering (i.e. $U = \mathbf{0}$). Thus,

$$x_k^0 = -(\xi_1 + \xi_2 + \ldots + \xi_k). \tag{6.2}$$

Let the stopping time

$$\tau_y := \inf\{k : \xi_1 + \xi_2 + \ldots + \xi_k \geq y\} \tag{6.3}$$

be the time required for the cumulative demand to reach y.

The following two lemmas are needed to establish that the cost up to τ_y for a policy that does not order grows faster than that of a policy that always orders the last period's demand. We show in Lemma 6.1 that the cost up to τ_y for the no-order policy grows faster than linear in y, while Lemma 6.2 states that ordering the last period's demand results in cost growing linearly in y.

LEMMA 6.1 *For $y \to \infty$ and any given real number d, the expression*

$$\frac{1}{y}\mathsf{E}\left(\sum_{k=0}^{\tau_y} f(i_k, -\sum_{t=1}^{k} \xi_t)\right)$$

is bounded from below by a function $\nu_d(y)$ which approaches d as $y \to \infty$.

Proof. Fix the initial state i. We define recursively the sequence $(t_k^j)_{k=0}^{\infty}$ of periods, in which the demand state is j, by

$$\begin{aligned} t_0^j &= 0, \\ t_{k+1}^j &= \inf\{n > t_k^j : i_n = j\}, \qquad k = 0, 1, \ldots. \end{aligned} \tag{6.4}$$

By convention, we define $t_\infty^j = \infty$. We refer to the interval containing the periods t_k^j through $t_{k+1}^j - 1$ as the kth cycle, $k = 0, 1, \ldots$.

The kth cycle demand and surplus cost, respectively, are

$$X_k^j := x_{t_{k+1}^j}^0 - x_{t_k^j}^0 = \xi_{t_k^j+1} + \ldots + \xi_{t_{k+1}^j}, \quad Y_k^j := \sum_{n=t_k^j}^{t_{k+1}^j-1} f(i_n, x_n^0). \tag{6.5}$$

Because of the Markov property of the demand process and the fact that $i_{t_k^j} = j$ for all $k \geq 1$, we know that the random variables X_k^j

are independent for $k \geq 0$, and identically distributed for $k = 1, 2, \ldots$. Furthermore, it is easy to prove that $\mathsf{E}X_0^j < \infty$ and $\mathsf{E}X_1^j < \infty$, and thus $\mathsf{E}X_k^j < \infty$, $k = 0, 1, \ldots$. Let

$$T_y^j := \inf\{n : x^0_{t_n^j} \leq -y\} = \inf\{n : \sum_{t=0}^{t_n^j} \xi_t \geq y\} = \inf\{n : \sum_{k=0}^{n-1} X_k \geq y\}$$

be the index of the first cycle for which the beginning shortage corresponding to the no-ordering policy is larger or equal to y. Thus, $t^j_{T_y^j}$ is the index of the first period in which the demand state is j and the surplus level at the beginning of the period in less than or equal to $-y$. On the other hand, τ_y is the index of the first period in which the starting surplus is less than or equal to $-y$, regardless of the demand state. Therefore, it is clear that

$$t^j_{T_y^j - 1} < \tau_y \leq t^j_{T_y^j}. \tag{6.6}$$

Also note that T_y^j is nondecreasing in y. By construction, $\{T_y^j, y \geq 0\}$ is a delayed (or general) renewal process with waiting times $\{X_k^j, k = 0, 1, \ldots\}$; (see, e.g., Ross (1983)).

Now consider the case where the demand state $j = l$, the special state defined in Assumption (iii). Because of Assumptions (ii) and (iii), we have $\lim_{x\to\infty} f(l, -x) = \infty$. Therefore, we can choose a constant y_d, $0 \leq y_d < \infty$, such that $f(l, -x) \geq \mathsf{E}(X_1^l)d$ for all $-x \leq -y_d$. Then, from (6.2), (6.5), and (6.6), we have

$$\begin{aligned}\sum_{k=0}^{\tau_y} f(i_k, -\sum_{t=1}^{k} \xi_t) \;&\geq\; \sum_{k=0}^{t_{T_y^l - 1}} f(i_k, x_k^0) \geq \sum_{k=0}^{T_y^l - 2} Y_k^l \geq \sum_{k=T^l_{y_d}}^{T_y^l - 2} Y_k^l \\ &\geq\; \sum_{k=T^l_{y_d}}^{T_y^l - 2} \mathsf{E}(X_1)d \geq (T_y^l - T_{y_d}^l - 1)\mathsf{E}(X_1^l)d,\end{aligned}$$

for any $y \geq y_d$. Clearly $\mathsf{E}(T^l_{y_d}) < \infty$, and the Elementary Renewal Theorem B.3.1 yields

$$\lim_{y\to\infty} \frac{\mathsf{E}(T_y^l)}{y} = \frac{1}{\mathsf{E}(X_1^l)}.$$

Finally with $\nu_d(y) := \mathsf{E}(T_y^l - T_{y_d}^l - 1)\mathsf{E}(X_1^l)d/y$, we obtain

$$\frac{1}{y}\mathsf{E}\left(\sum_{k=0}^{\tau_y} f(i_k, -\sum_{t=1}^{k} \xi_t)\right) \geq \nu_d(y) \quad \text{and} \quad \lim_{y\to\infty} \nu_d(y) = d.$$

This completes the proof. □

LEMMA 6.2 *For $y \to \infty$, the expression*

$$\frac{1}{y}\mathsf{E}\left(\sum_{k=0}^{\tau_y}(c(i_k,\xi_k)+f(i_k,-\xi_k))\right)$$

is bounded from above by a function $\bar{\nu}(y)$ which tends to a constant $c < \infty$ as $y \to \infty$.

Proof. Let X_k^j and T_y^j be defined as in the proof of Lemma 6.1. In what follows, we set the initial state $j := i_0 = i$, and omit the superscript. With this, the 0th cycle also begins with its first period in demand state i. Let

$$Z_k := \sum_{n=t_k+1}^{t_{k+1}} (c(i_n,\xi_n)+f(i_n,-\xi_n))$$

represent a 'reward' corresponding to the kth cycle. The stationarity of the Markovian demands implies that the pairs (X_k, Z_k) are i.i.d. for all k, including $k = 0$. Furthermore, we already know $\mathsf{E}(X_0) > 0$, and because of Assumptions (i), (ii), and (viii), we have $\mathsf{E}(Z_0) < \infty$.

Now we have

$$\begin{aligned}
&\frac{1}{y}\mathsf{E}\left(\sum_{k=0}^{\tau_y}(c(i_k,\xi_k)+f(i_k,-\xi_k))\right)\\
&\le \frac{1}{y}\mathsf{E}\left(\sum_{k=0}^{t_{T_y}}(c(i_k,\xi_k)+f(i_k,-\xi_k))\right)\\
&\le \frac{1}{y}\mathsf{E}\left(\sum_{k=0}^{T_y-1} Z_k\right) \le \frac{1}{y}\mathsf{E}\left(\sum_{k=0}^{T_y} Z_k\right). \qquad (6.7)
\end{aligned}$$

Setting the RHS of (6.7) equal to $\bar{\nu}(y)$, we conclude from Theorem B.4.1 that

$$\lim_{y\to\infty}\bar{\nu}(y) = \lim_{y\to\infty}\frac{1}{y}\mathsf{E}\left(\sum_{k=0}^{T_y} Z_k\right) = \frac{\mathsf{E}(Z_0)}{\mathsf{E}(X_0)} := c < \infty.$$

This completes the proof. □

LEMMA 6.3 *Under Assumptions* (i)-(iii) *and* (v)-(viii), *there are real numbers $0 < \alpha_0 < 1$ and $C_2 > -\infty$ such that $\max_{i\in I}\{s_i^\alpha\} \ge C_2$ for all $\alpha \in [\alpha_0, 1)$.*

Proof. Fix a discount factor α, $0 < \alpha < 1$. Let U be an optimal strategy with parameters (s_i^α, S_i^α). Let us fix a positive real number y and assume $s_i^\alpha < -y$ for each $i \in I$. In what follows, we specify a value of y, namely y^*, in terms of which we will construct an alternative strategy that is better than U, at least when the initial surplus $x = 0$.

Consider the alternate policy $\tilde{U} = (\tilde{u}_0, \tilde{u}_1, \ldots)$ defined by

$$\begin{cases} \tilde{u}_0 = 0, \\ \tilde{u}_k = \xi_k, \; k = 1, 2, \ldots, \tau_y - 1, \\ \tilde{u}_k = [x_k + u_k - \tilde{x}_k]^+, \; k \geq \tau_y, \end{cases}$$

where τ_y is as in (6.3).

We investigate the strategy $\tilde{U}$ with the initial surplus $\tilde{x}_0 = 0$. Let $\tilde{x}_k$ denote the inventory level in period k resulting from $\tilde{U}$. With the initial inventory level $x_0 = \tilde{x}_0 = 0$, we know that under the optimal policy U, no order is placed in or before the period τ_y since the corresponding surplus $x_k > -y > s_i^\alpha$, $k = 0, 1, \ldots, \tau_y - 1$. It is easy to see that

$$\tilde{x}_k = -\xi_k \geq x_k = -\sum_{t=1}^{k} \xi_t, \quad k \leq \tau_y - 1.$$

For $k \geq \tau_y$, we have the following situation. As long as the inventory level after ordering for U is less than the inventory level before ordering for $\tilde{U}$, the strategy $\tilde{U}$ orders nothing. As soon as $x_k + u_k$ is greater than or equal to $\tilde{x}_k$, we order up to the level $x_k + u_k$, and the inventory level is the same for both strategies from then on. Obviously, the expected cost corresponding to U from τ_y on is not smaller than the one corresponding to $\tilde{U}$ from τ_y on. Therefore,

$$\begin{aligned} & v^\alpha(i, 0) - J^\alpha(i, 0, \tilde{U}) \\ &= \mathsf{E}\left(\sum_{k=0}^{\infty} \alpha^k (c(i_k, u_k) + f(i_k, x_k) - c(i_k, \tilde{u}_k) - f(i_k, \tilde{x}_k)) \right) \\ &\geq \mathsf{E}\left(\sum_{k=0}^{\tau_y - 1} \alpha^k (f(i_k, -\sum_{t=1}^{k} \xi_t) - c(i_k, \xi_k) - f(i_k, -\xi_k)) \right). \end{aligned} \tag{6.8}$$

Let

$$g(y, \alpha) := \mathsf{E}\left(\sum_{k=0}^{\tau_y - 1} \alpha^k (f(i_k, -\sum_{t=1}^{k} \xi_t) - c(i_k, \xi_k) - f(i_k, -\xi_k)) \right).$$

It follows from Lemmas 6.2 and 6.1 that there is a $y^* > 0$ such that

$$\begin{aligned} g(y^*,1) &= \mathsf{E}\left(\sum_{k=0}^{\tau_{y^*}-1} f(i_k, -\sum_{t=1}^{k}\xi_t)\right) \\ &\quad -\mathsf{E}\left(\sum_{k=0}^{\tau_{y^*}-1}(c(i_k,\xi_k)+f(i_k,-\xi_k))\right) > 0. \end{aligned}$$

Because $\lim_{\alpha\to 1} g(y^*,\alpha) = g(y^*,1)$, it follows that there is an $\alpha_0 < 1$ such that for all $\alpha \geq \alpha_0$, we have $g(y^*,\alpha) > 0$. Therefore,

$$v^\alpha(i,0) - J^\alpha(i,0,\tilde{U}) > 0, \quad 1 > \alpha \geq \alpha_0,$$

for the policy $\tilde{U}$ defined by $y = y^*$. This leads to a contradiction that the policy U is optimal, which helps us to conclude that any strategy with $\max\{s_i^\alpha\} < C_2 = -y^*$ cannot be optimal. □

LEMMA 6.4 *Let α_0 be the same as in Lemma* 6.3. *If Assumptions* (i)-(iii) *and* (v)-(viii) *are satisfied, then $v^\alpha(i,\cdot)$ is locally equi-Lipschitzian for $\alpha \geq \alpha_0$, i.e., for any $b > 0$, there exists a positive constant $C_5 < \infty$, independent of α, such that*

$$|v^\alpha(i,x) - v^\alpha(i,\tilde{x})| \leq C_5|x-\tilde{x}| \quad \forall \quad x,\tilde{x} \in [-b,b]. \tag{6.9}$$

Proof. Note that from the convexity and polynomial growth properties of $f(i,\cdot)$, it follows that $f(i,\cdot)$ is locally Lipschitz continuous. The proof is very similar to the one of Lemma 5.3.

Consider the case $\tilde{x} \geq x$. Let us fix an $\alpha \geq \alpha_0$. Then use the optimal strategy U with the initial surplus x, and the strategy $\tilde{U}$ defined by

$$\tilde{u}_k = [u_k - (\tilde{x}_k - x_k)]^+ = \begin{cases} 0 & \text{if } \ u_k \leq \tilde{x}_k - x_k, \\ u_k + x_k - \tilde{x}_k & \text{if } \ u_k > \tilde{x}_k - x_k, \end{cases}$$

with the initial surplus $\tilde{x}$. (Here x_k and $\tilde{x}_k$ are the surplus levels at time k under the strategies U and $\tilde{U}$, respectively). It is easy to see that if we define $\tilde{\tau} := \inf\{t : \tilde{u}_t > 0\}$, then

$$\tilde{u}_k = \begin{cases} 0 & \text{for } \ 0 \leq k \leq \tilde{\tau}-1, \\ u_k + x_k - \tilde{x}_k & \text{for } \ k = \tilde{\tau}, \\ u_k & \text{for } \ k \geq \tilde{\tau}+1, \end{cases} \tag{6.10}$$

and, therefore,

$$\tilde{x}_{\tilde{\tau}+1} = \tilde{x}_{\tilde{\tau}} + \tilde{u}_{\tilde{\tau}} - \xi_{\tilde{\tau}} = x_{\tilde{\tau}} + u_{\tilde{\tau}} - \xi_{\tilde{\tau}} = x_{\tilde{\tau}+1}.$$

It follows that for $k > \tilde{\tau}$, the surplus levels $\tilde{x}_k$ and x_k are equal and the order quantities $\tilde{u}_k$ and u_k are equal.

Furthermore, it is easy to conclude that

$$0 \le \tilde{u}_k \le u_k \text{ and } |\tilde{x}_k - x_k| \le |\tilde{x} - x| \text{ for all } k.$$

From Assumptions (i) and (ii), we have

$$c(i_k, \tilde{u}_k) \le c(i_k, u_k) \text{ and } |f(i_k, \tilde{x}_k) - f(i_k, x_k)| \le C|\tilde{x} - x|.$$

Therefore,

$$\begin{aligned} & v^\alpha(i, \tilde{x}) - v^\alpha(i, x) \\ & \quad \le \; J^\alpha(i, \tilde{x}; \tilde{U}) - J^\alpha(i, x; U) \\ & \quad = \; \mathsf{E}\Big(\sum_{k=0}^{\tilde{\tau}} \alpha^k (f(i_k, \tilde{x}_k) - f(i_k, x_k) + c(i_k, \tilde{u}_k) - c(i_k, u_k))\Big) \\ & \quad \le \; \mathsf{E}\Big(\sum_{k=0}^{\tilde{\tau}} \alpha^k C|\tilde{x} - x|\Big) \le \mathsf{E}(\tilde{\tau} + 1) C|\tilde{x} - x|. \end{aligned} \tag{6.11}$$

In order to prove that $\mathsf{E}(\tilde{\tau} + 1)$ is bounded, note from (6.10) that $\tilde{x}_k - x_k = \tilde{x} - x - \sum_{t=0}^{k-1} u_k$ for $k \le \tilde{\tau}$ and, therefore,

$$\tilde{\tau} = \inf\{k : u_k > \tilde{x}_k - x_k\} = \inf\{k : \sum_{t=0}^{k} u_k > \tilde{x} - x\}.$$

Let i^* be such that $s_{i^*}^\alpha \ge C_2$. Consider the inventory level $y_n = x_n + u_n$ after ordering in period n, if there exists one, in which $i_n = i^*$. From the structure of the optimal strategy, we know that under the assumption $\alpha \ge \alpha_0$ we have $y_n \ge C_2$ (see Lemma 6.3), and therefore,

$$C_2 - 1 < y_n = x + \sum_{j=0}^{n} u_j - \sum_{j=0}^{n-1} \xi_j.$$

This implies that if $\sum_{j=0}^{n-1} \xi_j > \tilde{x} - C_2 + 1$ and $i_n = i^*$, then $\sum_{j=0}^{n} u_j > \tilde{x} - x$, and therefore $\tilde{\tau} \le n$, a.s. It is easy to see from Lemma 5.1, the fact that $x, \tilde{x} \in [-b, b]$, and Assumption (vi) that

$$\begin{aligned} \mathsf{E}(\tilde{\tau}) \; & \le \; \mathsf{E}[\inf\{k : \sum_{j=0}^{k-1} \xi_j \ge \tilde{x} - C_2 + 1 \text{ and } i_k = i^*\}] \\ & \le \; \mathsf{E}_i(\tau_{\tilde{x} - C_2 + 1} + \inf\{k \ge \tau_{\tilde{x} - C_2 + 1} : i_k = i^*\}) < \infty, \end{aligned}$$

where τ_y is defined by (6.3). This inequality, together with (6.11), implies that for $\alpha \geq \alpha_0$ and $\tilde{x} \geq x$, we have

$$v^\alpha(i,\tilde{x}) - v^\alpha(i,x) \leq C_5|\tilde{x} - x|.$$

To complete the proof, it is sufficient to prove the above inequality for $\tilde{x} < x$. In this case, let us define

$$\tilde{u}_k = \begin{cases} u_k + x - \tilde{x} & \text{if } k = k^*, \\ u_k & \text{otherwise,} \end{cases}$$

where k^* is the first k with $u_k > 0$. It is obvious that

$$\begin{aligned} & v^\alpha(i,\tilde{x}) - v^\alpha(i,x) \\ \leq\ & J^\alpha(i,\tilde{x};\tilde{U}) - J^\alpha(i,x;U) \\ =\ & \mathsf{E}\Big(\sum_{k=0}^{k^*} \alpha^k (f(i_k,\tilde{x}_k) - f(i_k,x_k)) + \alpha^{k^*} c_{i_{k^*}}(\tilde{u}_{k^*} - u_{k^*})\Big) \\ \leq\ & \mathsf{E}(k^*+1)C|\tilde{x} - x| + \max\{c_i : i \in \mathbb{I}\}|\tilde{x} - x| \\ \leq\ & C_5|\tilde{x} - x|, \end{aligned} \tag{6.12}$$

where the boundedness of $\mathsf{E}(k^*+1)$ can be proved in a similar manner as above. □

6.3. Vanishing Discount Approach

THEOREM 6.1 *Under Assumptions* (i)-(iii) *and* (v)-(viii), *the following results hold.*

(a) *Let α_0 be the same as in Lemma* 6.3. *Then, the differential discounted value function $w^\alpha(i,x) := v^\alpha(i,x) - v^\alpha(1,0)$ is uniformly bounded with respect to α, $\alpha \geq \alpha_0$, for each fixed x and i, and $(1-\alpha)v^\alpha(1,0)$ is uniformly bounded on $0 < \alpha < 1$.*

(b) *There exist a sequence $(\alpha_k)_{k=1}^\infty$ converging to one, a constant λ^*, and a locally Lipschitz continuous function $w^*(\cdot,\cdot)$, such that*

$$(1-\alpha_k)v^{\alpha_k}(i,x) \to \lambda^* \quad \textit{and} \quad w^{\alpha_k}(i,x) \to w^*(i,x), \tag{6.13}$$

locally uniformly in x and i as k goes to infinity.

(c) *$w^*(i,\cdot)$ is K-convex.*

Proof. To prove (a) we first show that the differential discounted value function $w^\alpha(i,x) := v^\alpha(i,x) - v^\alpha(1,0)$ is uniformly bounded with respect to α, $\alpha \geq \alpha_0$ for each fixed x and i. The proof of this part is similar to the proof of Lemma 5.5.

Lemma 6.4 implies

$$\begin{aligned} |w^\alpha(i,x)| &= |v^\alpha(i,x) - v^\alpha(1,0)| \\ &\le |v^\alpha(i,x) - v^\alpha(i,0)| + |v^\alpha(i,0) - v^\alpha(1,0)| \\ &\le C_5|x| + |w^\alpha(i,0)|. \end{aligned}$$

Thus, it is sufficient to prove that $w^\alpha(i,0)$ is uniformly bounded. (Note that C_5 may depend on x, but it is independent of $\alpha \ge \alpha_0$.)

First, we show that there is an $\underline{M} > -\infty$ with $w^\alpha(i,0) \ge \underline{M}$ for all $\alpha \ge \alpha_0$. Let α be fixed. We have shown in Section 3.6 that in this discounted case there is a stationary optimal feedback policy $U = (u(i,x), u(i,x), \ldots)$. With $k^* = \inf\{k : i_k = i\}$, we consider the cost for the initial state $(i_0, \tilde{x}_0) = (1,0)$ and the following inventory policy $\tilde{U}$:

$$\tilde{u}_k = \xi_{k-1} \text{ for } k < k^*,\ \tilde{u}_{k^*} = u(i_{k^*},0) - x_{k^*} = u(i_{k^*},0) + \xi_{k^*-1},$$
$$\text{and } \tilde{u}_k = u(i_k, \tilde{x}_k) \text{ for } k > k^*.$$

That is, we order the previous period's demand until the demand state is equal to i in period k^*. In that period we order as much as is needed to reach an inventory level after ordering, that is equal to the inventory level in period 0 after ordering the optimal amount $u(i,0)$ for the problem starting in the state $(i,0)$. From that instant on, we apply the optimal feedback policy U.

The cost corresponding to this policy is

$$\begin{aligned} &J^\alpha(1,0;\tilde{U}) \\ &= \mathsf{E}\left(\sum_{k=0}^{k^*-1} \alpha^k (c(i_k,\tilde{u}_k) + f(i_k,\tilde{x}_k)) + \alpha^{k^*} v^\alpha(i, -\xi_{k^*-1})\right). \qquad (6.14) \end{aligned}$$

To derive an upper bound for $\mathsf{E}(v^\alpha(i,-\xi_{k^*-1}))$ in terms of $v^\alpha(i,0)$, we use the dynamic programming equation (3.51). For $y \ge 0$, we have

$$v^\alpha(i,-y) \le f(i,-y) + c(i,u(i,0)+y) + \alpha \sum_{j=1}^{L} p_{ij} \int_0^\infty v^\alpha(j, u(i,0)-z)\varphi_i(z)dz.$$

On the other hand, because U is optimal, we have

$$v^\alpha(i,0) = f(i,0) + c(i,u(i,0)) + \alpha \sum_{j=1}^{L} p_{ij} \int_0^\infty v^\alpha(j, u(i,0) - z)\varphi_i(z)dz.$$

Thus,

$$\begin{aligned} v^\alpha(i,-y) \le\ & v^\alpha(i,0) + f(i,-y) + c(i,u(i,0)+y) \\ & - f(i,0) - c(i,u(i,0)). \qquad (6.15) \end{aligned}$$

We use (6.15) with $y = \xi_{k^*-1}$, use $\tilde{u}_k = \xi_{k-1}$ and $\tilde{x}_k = -\xi_{k-1}$ for $k < k^*$ in (6.14), and combine them to obtain

$$\begin{aligned} J^\alpha(1,0;\tilde{U}) &\le \mathsf{E}\Big(\sum_{k=0}^{k^*-1} \alpha^k (c(i_k, \xi_{k-1}) + f(i_k, -\xi_{k-1})) + \alpha^{k^*} v^\alpha(i,0) \\ &\quad + \alpha^{k^*} (c(i, u(i,0) + \xi_{k^*-1}) + f(i, -\xi_{k^*-1}) \\ &\quad - c(i, u(i,0)) - f(i,0))\Big). \end{aligned} \tag{6.16}$$

Because of Assumptions (i), (ii), and (viii), all arguments of the sums in the first line and the expressions on the second line are uniformly bounded from above by a constant $\underline{M}$. Moreover, $\mathsf{E}(k^*) < \infty$ because of Assumption (vi). Therefore, it holds that

$$\begin{aligned} w^\alpha(i,0) &= v^\alpha(i,0) - v^\alpha(1,0) \ge v^\alpha(i,0) - J^\alpha(1,0;\tilde{U}) \\ &\ge v^\alpha(i,0) - \mathsf{E}\Big(\sum_{k=0}^{k^*-1} \alpha^k (c(i_k, \xi_{k-1}) + f(i_k, -\xi_{k-1})) \\ &\quad + \alpha^{k^*} v^\alpha(i,0) + \alpha^{k^*} (c(i, u(i,0) + \xi_{k^*-1}) \\ &\quad + f(i, -\xi_{k^*-1}) - c(i, u(i,0)) - f(i,0))\Big) \\ &\ge v^\alpha(i,0)(1 - \mathsf{E}(\alpha^{k^*})) + \underline{M} \ge \underline{M}. \end{aligned} \tag{6.17}$$

The opposite inequality $w^\alpha(i,0) \le \overline{M}$ is shown analogously by changing the role of the states 1 and i. Thus, $|w^\alpha(i,x)| \le C_5|x| + \max\{\underline{M}, \overline{M}\}$, which proves that $w^\alpha(i,x)$ is uniformly bounded with respect to α, $\alpha \ge \alpha_0$.

We now prove the second part of (a), i.e., that $(1-\alpha)v^\alpha(1,0)$ is uniformly bounded on $0 < \alpha < 1$. Consider the strategy $\tilde{U} = (0, \xi_0, \xi_1, \ldots)$. Then, because $\tilde{U}$ is not necessarily optimal,

$$0 \le v^\alpha(1,0) \le J^\alpha(1,0;\tilde{U}) = \mathsf{E}\left(\sum_{k=0}^{\infty} \alpha^k (c(i_k, \xi_{k-1}) + f(i_k, -\xi_{k-1})) \right).$$

Owing to Assumptions (i), (ii), and (viii), there exists a constant $C_9 < \infty$ such that $\mathsf{E}(c(i_k, \xi_{k-1}) + f(i_k, -\xi_{k-1})) < C_9$. Therefore,

$$\begin{aligned} 0 \le (1-\alpha)v^\alpha(1,0) &\le (1-\alpha) \sum_{k=0}^{\infty} \alpha^k \mathsf{E}(c(i_k, \xi_{k-1}) + f(i_k, -\xi_{k-1})) \\ &\le (1-\alpha) \sum_{k=0}^{\infty} \alpha^k C_9 = C_9. \end{aligned}$$

This completes the proof of part (a) of the theorem.

It is immediate from Lemma 6.4 and the definition of $w^\alpha(i,x)$ that $w^\alpha(i,\cdot)$ is locally equi-Lipschitzian for $\alpha \geq \alpha_0$, and therefore it is uniformly equicontinuous in any finite interval by Theorem A.3.2. Additionally, according to part (a) of Theorem 6.1, $w^\alpha(i,\cdot)$ and $(1-\alpha)v^\alpha(1,0)$ are uniformly bounded. Therefore, from the Arzelà-Ascoli Theorem A.3.5 and Lemma 6.4, there is a sequence $\alpha_k \to 1$, a locally Lipschitz continuous function $w^*(i,x)$, and a constant λ^* such that

$$(1-\alpha_k)v^{\alpha_k}(1,0) \to \lambda^*$$

and $w^{\alpha_k}(i,x)$ is locally uniformly convergent to $w^*(i,x)$. This completes the proof of part (b).

The proof of (c) follows from the fact that w^* as a limit of K-convex functions is K-convex; (see Theorem C.2.2). □

LEMMA 6.5 *There are constants $\alpha_2 \in [0,1)$ and $C_8 > 0$ such that for all $\alpha \geq \alpha_2$, we have $S_i^\alpha \leq C_8 < \infty$ for any i for which $s_i^\alpha > -\infty$.*

Proof. Let us fix the initial state $i_0 = i$ for which $s_i^\alpha > -\infty$. Fix $\alpha_2 > 0$ and a discount factor $\alpha \geq \alpha_2$. Let $U = (u(i_0,x_0), u(i_1,x_1),\ldots)$ be an optimal strategy with parameters (s_j^α, S_j^α), $j \in I$. Let us fix a positive real number y and assume $S_i^\alpha > y$. In what follows, we specify a value of y, namely y^*, in terms of which we will construct an alternative strategy $\tilde{U}$ that is better than U.

For the demand state g specified in Assumption (iv), let

$$\tau^g := \inf\{n > 0 : i_n = g\}$$

be the first period (not counting the period 0) with the demand state g. Furthermore, let d be the state with the lowest per unit ordering cost, i.e., $c_d \leq c_i$ for all $i \in I$. Then, we define

$$\tau := \inf\{n \geq \tau^g : i_n = d\}.$$

Assume $x_0 = \tilde{x}_0 = \bar{x} := \min\{0, s_i^\alpha\}$ and consider the policy $\tilde{U}$ defined by

$$\begin{cases} \tilde{u}_0 = -\bar{x} \geq 0, \\ \tilde{u}_k = 0,\ k = 1,2,\ldots,\tau-1, \\ \tilde{u}_\tau = x_\tau + u(i_\tau, x_\tau) - \tilde{x}_\tau \geq 0, \\ \tilde{u}_k = u(i_k,x_k),\ k \geq \tau+1. \end{cases}$$

The two policies and the resulting trajectories differ only in the periods 0 through τ. Therefore, we have

$$v^\alpha(i,\bar{x}) - J^\alpha(i,\bar{x};\tilde{U})$$

$$
\begin{aligned}
&= J^\alpha(i,\bar{x};U) - J^\alpha(i,\bar{x};\tilde{U}) \\
&= \mathsf{E}\left(\sum_{k=0}^{\tau} \alpha^k (f(i_k,x_k) - f(i_k,\tilde{x}_k) + c(i_k,u_k) - c(i_k,\tilde{u}_k))\right) \\
&= \mathsf{E}\left(\sum_{k=1}^{\tau} \alpha^k (f(i_k,x_k) - f(i_k,-\sum_{t=1}^{k}\xi_k))\right) \\
&\quad +\mathsf{E}\left(\sum_{k=0}^{\tau} \alpha^k c(i_k,u_k)\right) - \mathsf{E}(\alpha^\tau c(i_\tau,\tilde{u}_\tau)) - c(i,-\bar{x}). \qquad (6.18)
\end{aligned}
$$

After ordering in period τ, the total accumulated ordered amount up to period τ is the same for both the policies, because $x_0 = \tilde{x}_0$ and $x_{\tau+1} = \tilde{x}_{\tau+1}$. Observe that the policy $\tilde{U}$ orders only in the periods 0 and τ. In the period 0, $\tilde{u}_0 \leq u_0$ so that $\tilde{U}$ orders no more than U at the unit cost c_i. The second order of the policy $\tilde{U}$ is executed at the lowest possible per unit cost c_d in period τ which is not earlier than any of the ordering periods of policy U. Because $\tilde{U}$ orders at most twice in periods $0, 1, \ldots, \tau$, the total fixed ordering cost of $\tilde{U}$ exceeds the total fixed ordering cost of U by at most $2K$. Thus,

$$
2K + \mathsf{E}\left(\sum_{k=0}^{\tau} \alpha^k c(i_k,u_k)\right) \geq c(i,-\bar{x}) + \mathsf{E}(\alpha^\tau c(i_\tau,\tilde{u}_\tau)).
$$

Furthermore, it follows from Assumptions (ii) and (vi)–(viii) that

$$
\mathsf{E}\left(\sum_{k=1}^{\tau} f(i_k,-\sum_{t=1}^{k}\xi_t)\right) < \infty.
$$

Because $\tau \geq \tau^g$, we obtain

$$
\mathsf{E}\left(\sum_{k=1}^{\tau} \alpha^k f(i_k,x_k)\right) \geq \mathsf{E}\left(\sum_{k=1}^{\tau^g} \alpha^k f(i_k,x_k)\right),
$$

and because $s_i^\alpha \geq y$, we obtain

$$
x_k \geq y - \sum_{t=1}^{k} \xi_t.
$$

Irreducibility of the Markov chain $(i_n)_{n=0}^{\infty}$ implies the existence of an integer m, $0 \leq m \leq L$, such that $\mathsf{P}(i_m = g) > 0$. Let m_0 be the smallest of such m. It follows that $\tau^g \geq m_0$ and therefore,

$$\begin{aligned} & \mathsf{E}\left(\sum_{k=1}^{\tau^g} \alpha^k f(i_k, x_k)\right) \\ \geq \ & \alpha^{m_0} \mathsf{E}(f(i_{m_0}, x_{m_0})) \\ \geq \ & \alpha_2^{m_0} \mathsf{E}(f(g, y - \sum_{t=1}^{m_0} \xi_t) | i_{m_0} = g) \mathsf{P}(i_{m_0} = g), \ \forall \alpha \geq \alpha_2. \quad (6.19) \end{aligned}$$

Using Assumptions (ii), (iv), and (viii), it is easy to show that the RHS of (6.19) (and consequently of (6.18)) tends to infinity as y goes to infinity. Therefore, we can choose y^*, $0 \leq y^* < \infty$, such that for all $\alpha \geq \alpha_2$,

$$v^\alpha(i, \bar{x}) - J^\alpha(i, \bar{x}; \tilde{U}) \geq \mathsf{E}\left(\sum_{k=1}^{\tau} \alpha^k (f(i_k, x_k) - f(i_k, -\sum_{t=1}^{k} \xi_k))\right) - 2K > 0.$$

Thus, for $\alpha \geq \alpha_2$, a policy with $S_i^\alpha > C_8 := y^*$ cannot be optimal. □

LEMMA 6.6 *For the sequence* $\{\alpha_k\}$ *defined in Theorem* 6.1, *there is a positive integer* N_0 *such that for* $k \geq N_0$, *functions* $w^{\alpha_k}(i, \cdot)$ *on* $(-\infty, 0]$ *are uniformly bounded from above by a function of polynomial growth with power* γ.

Proof. Choose N_1 such that $\alpha_k > \alpha_2$ for $k \geq N_1$, with α_2 defined in Lemma 6.5. Substituting $v^{\alpha_k}(i, x) = w^{\alpha_k}(i, x) + v^{\alpha_k}(1, 0)$ in (3.51) yields

$$\begin{aligned} & w^{\alpha_k}(i, x) + (1 - \alpha_k) v^{\alpha_k}(1, 0) \\ & = \ f(i, x) + \inf_{u \geq 0} \{c(i, u) + \alpha_k F(w^{\alpha_k})(i, x + u)\}. \quad (6.20) \end{aligned}$$

First, let $x \geq C_8$, where C_8 is defined in Lemma 6.5. Then it follows from the structure of the optimal policy and Lemma 6.5 that the infimum is obtained for $u = 0$, and we find

$$w^{\alpha_k}(i, x) + (1 - \alpha_k) v^{\alpha_k}(1, 0) = f(i, x) + \alpha_k F(w^{\alpha_k})(i, x).$$

Because $\lim_{k \to \infty} (1 - \alpha_k) v^{\alpha_k}(1, 0) = \lambda^*$ and $\lim_{k \to \infty} w^{\alpha_k}(i, x) = w^*(i, x)$, it follows that

$$\lim_{k \to \infty} \alpha_k F(w^{\alpha_k})(i, x) = w^*(i, x) + \lambda^* - f(i, x).$$

Now let $x < C_8$ and set $u = C_8 - x > 0$. Then, because the infimum is not necessarily attained at u, we have

$$w^{\alpha_k}(i, x) + (1 - \alpha_k) v^{\alpha_k}(1, 0) \leq f(i, x) + c(i, C_8 - x) + \alpha_k F(w^{\alpha_k})(i, C_8).$$

Choose N_0, $N_0 \geq N_1$, such that for all $k \geq N_0$,

$$|(1-\alpha_k)v^{\alpha_k}(1,0) - \lambda^*| \leq \varepsilon$$

and

$$|\alpha_k F(w^{\alpha_k})(i, C_8) - w^*(i, C_8) + \lambda^* - f(i, C_8)| \leq \varepsilon.$$

Then for $k \geq N_0$, we obtain

$$w^{\alpha_k}(i,x) \leq f(i,x) + c(i, C_8 - x) + w^*(i, C_8) + \lambda^* - f(i, C_8) + 2\varepsilon,$$

which provides a uniform upper bound of growth rate γ for $x \leq 0$. □

LEMMA 6.7 *For α_0 obtained in Lemma 6.3, there is a finite constant C_9 such that for all α, $\alpha_0 \leq \alpha < 1$,*

$$w^{\alpha}(i,x) \geq \begin{cases} C_9 - c_i x, & x > 0, \\ C_9, & x \leq 0. \end{cases}$$

Proof. First assume $x > 0$. We know that there is an optimal stationary feedback policy $U = (u(i_0, x_0), u(i_1, x_1), \ldots)$. Consider the modified policy $\tilde{U} = (u(i_0, x_0) + x, u(i_1, x_1), u(i_2, x_2), \ldots)$. If we use U with the initial inventory $x_0 = x$ and $\tilde{U}$ with the initial inventory $\tilde{x}_0 = 0$, then under both policies the trajectories differ only in the initial period. Therefore, we have

$$\begin{aligned} w^{\alpha}(i,x) &= v^{\alpha}(i,x) - v^{\alpha}(i,0) + w^{\alpha}(i,0) \\ &\geq v^{\alpha}(i,x) - J^{\alpha}(i,0;\tilde{U}) + w^{\alpha}(i,0) \\ &= J^{\alpha}(i,x;U) - J^{\alpha}(i,0;\tilde{U}) + w^{\alpha}(i,0) \\ &= c(i,u) - c(i,\tilde{u}) + f(i,x) - f(i,0) + w^{\alpha}(i,0) \\ &\geq -K - c_i x + w^{\alpha}(i,0). \end{aligned} \tag{6.21}$$

By Lemma 6.3, $w^{\alpha}(i,0)$ is uniformly bounded for $\alpha \geq \alpha_0$. Therefore, there is a constant C_9 such that

$$w^{\alpha}(i,x) \geq C_9 - c_i x \text{ for } \alpha \geq \alpha_0 \text{ and } x > 0.$$

If $x \leq 0$, it follows from Lemma 5.4 that $w^{\alpha}(i,x)$ is decreasing in x. Then, by using the same arguments concerning the uniform boundedness of $w^{\alpha}(i,0)$ as above, we obtain

$$w^{\alpha}(i,x) \geq w^{\alpha}(i,0) \geq C_9,\ \alpha \geq \alpha_0 \text{ and } x \leq 0.$$

This completes the proof. □

THEOREM 6.2 *The limit* (λ^*, w^*), *obtained in Theorem* 6.1, *satisfies the average cost optimality equation*

$$w(i,x) + \lambda = f(i,x) + \inf_{u \geq 0} \{c(i,u) + F(w)(i,x+u)\}. \tag{6.22}$$

Furthermore, $w^* \in \mathbb{L}^{\gamma^-}$. *Moreover, let* U^α *be an optimal policy for the discounted cost problem for which the infimum in* (3.51) *is attained. Let* $u^*(i,x)$ *be a cluster point of* $(u^{\alpha_k}(i,x))_{k=0}^\infty$, *where the sequence* $(\alpha_k)_{k=1}^\infty$ *is defined in Theorem* 6.1. *Then the infimum in* (6.22) *is attained at* $u^*(i,x)$ *for* $w = w^*$.

Proof. From (6.20), it follows that for any u,

$$w^{\alpha_k}(i,x) + (1-\alpha_k)v^{\alpha_k}(1,0) \leq f(i,x) + c(i,u) + F(w^{\alpha_k})(i,x+u). \tag{6.23}$$

Now define the random variable

$$\eta_k(j,z) := w^{\alpha_k}(j, z - \xi^j),$$

where ξ^j is a random variable with the distribution $\mathsf{P}(\xi^j < t) = \Phi_j(t)$. Then, we can write

$$F(w^{\alpha_k})(i,z) = \sum_{j=1}^{L} p_{ij} \mathsf{E}(\eta_k(j,z)).$$

For α_k sufficiently close to one, it follows from Lemma 6.6 that there is a constant C_3 such that

$$\eta_k(j,z) \leq C_3(1 + |z - \xi^j|)^\gamma \leq C_4(1+\xi^j)^\gamma, \tag{6.24}$$

and it follows from Lemma 6.7 that

$$\eta_k(j,z) \geq \inf_{x \leq z} w^{\alpha_k}(j,x) = C_6 > -\infty, \tag{6.25}$$

where C_4 and C_6 depend on z, which is fixed. We know from Theorem 6.1 that w^{α_k} converges locally uniformly to w^*. Therefore, $\eta_k(j,z)$ converges everywhere to $\eta(j,z) := w^*(j, z - \xi^j)$. On the other hand, we see from (6.24) and (6.25) that

$$\mathsf{E}(|\eta_k(j,z)|)^{(\gamma+\varepsilon)/\gamma} \leq (\max\{C_4, C_6\})^{(\gamma+\varepsilon)/\gamma}(1+\xi^j)^{\gamma+\varepsilon} < \infty$$

by Assumption (viii). Therefore, the random variables $\eta_k(j,y,z)$ are uniformly integrable with respect to k for any fixed j and z, and we can pass to the limit in (6.23) to obtain, for each x and u,

$$w^*(i,x) + \lambda \leq f(i,x) + c(i,u) + F(w^*)(i,x+u). \tag{6.26}$$

Now let the infimum in (6.20) be attained at u_k for fixed x and j. Then, we have

$$w^{\alpha_k}(i,x)+(1-\alpha_k)v^{\alpha_k}(1,0) = f(i,x)+c(i,u_k)+F(w^{\alpha_k})(i,x+u_k). \quad (6.27)$$

It follows from Lemma 6.5 that, for sufficiently large k, u_k are uniformly bounded. Therefore, we can choose a subsequence (still denoted by $(\alpha_k)_{k=0}^{\infty}$) such that $u_k \to u^*$. Let

$$\hat{\eta}_k(j,x) := w^{\alpha_k}(j, x+u_k-\xi^j).$$

The boundedness of u_k also allows us to to prove the inequalities

$$\hat{\eta}_k(j,x) \leq C_3(1+|x+u_k-\xi^j|)^\gamma \leq \hat{C}_4(1+\xi^j)^\gamma \quad (6.28)$$

and

$$\hat{\eta}_k(j,x) \geq \inf_{v\leq x+u_k} w^{\alpha_k}(j,v) = \hat{C}_6 > -\infty. \quad (6.29)$$

It is easy to see that as $k \to \infty$,

$$\hat{\eta}_k(j,x) \to \hat{\eta}(j,x) := w^*(j, x+u^*-\xi^j).$$

The inequalities (6.28) and (6.29) imply that $\hat{\eta}_k(j,x)$ is uniformly integrable. By taking the limit in (6.27) as $k \to \infty$, we obtain

$$w^*(i,x)+\lambda \geq f(i,x)+c(i,u^*)+F(w^*)(i,x+u^*); \quad (6.30)$$

(see Section B.1 for details). The reason we write (6.30) as an inequality rather than an equality is because of the lower semicontinuity of $c(i,\cdot)$. Nevertheless, the inequality (6.30) together with (6.26) prove (6.22) and the last part of the theorem. The result that $w^* \in \mathbb{L}^{\gamma^-}$ immediately follows from Lemmas 6.6 and 6.7. □

LEMMA 6.8 *For each $i \in I$, the function $w^*(i,x)$ is decreasing for $x \leq 0$, $w^*(i,\cdot)$ is bounded from below, and* $\lim_{x\to\infty} w^*(i,x) = \infty$.

Proof. It follows from Lemma 5.4 that the function $w^\alpha(i,x)$ is decreasing for $x \leq 0$. This implies the same property for the limit w^*.

Because $w^*(i,x)$ is locally Lipschitz by Theorem 6.1, it follows that $w^*(i,0)$ is finite. Assume

$$\lim_{u\to\infty} (c(i,u)+F(w^*)(i,u)) = -\infty.$$

Then, it follows from (6.22) that $w^*(i,0) = -\infty$, and we obtain a contradiction. Therefore, $c(i,u)+F(w^*)(i,u)$ is bounded from below, as is $w^*(i,x)$.

Let $I_\infty := \{i \in I : \lim_{x\to\infty} w^*(i,x) = \infty\}$. Since $c(g,u) + F(w^*)(g,u)$ is bounded from below and Assumptions (i) and (iv) imply that $\lim_{x\to\infty} f(g,x) = \infty$, it follows that $\lim_{x\to\infty} w^*(g,x) = \infty$. Therefore, $g \in I_\infty \neq \emptyset$. Assume $I_\infty \neq I$ and choose $j \in I_\infty$ and $i \in I \setminus I_\infty$. Then, it follows that $p_{ij} = 0$. Otherwise, we would have $\lim_{x\to\infty} F(w^*)(i,x) = \infty$ and therefore, $\lim_{x\to\infty} w^*(i,x) = \infty$, which contradicts $i \notin I_\infty$. We can now conclude that $p_{ij} = 0$ for all $i \notin I_\infty$ and $j \in I_\infty$. The latter implies that the Markov process $(i_n)_{n=0}^\infty$ is not irreducible, which contradicts Assumption (vi). This proves $I_\infty = I$. □

6.4. Verification Theorem

DEFINITION 6.1 An admissible strategy $U = (u_0, u_1, \ldots)$ is called stable with respect to w if for each initial surplus level x and for each $i \in I$,

$$\lim_{k\to\infty} \frac{1}{k} \mathsf{E}(w(i_k, x_k)) = 0,$$

where x_k is the surplus level in period k corresponding to the initial state (i,x) and the strategy U.

LEMMA 6.9 *Let Assumptions* (i)-(viii) *hold. Let C_2 be the constant specified in Lemma* 6.3. *Then there are constants $S_i < \infty$ and $s_i \le S_i$, $i \in I$, satisfying* $\max_{i\in I}\{s_i\} \ge C_2$, *such that*

$$u^*(i,x) = \begin{cases} S_i - x, & x \le s_i, \\ 0, & x > s_i, \end{cases}$$

attains the minimum in the RHS *of* (6.22) *for $w = w^*$, as defined in Theorem* 6.1. *Furthermore, the stationary feedback strategy $U^* = (u^*, u^*, \ldots)$ is stable with respect to any $w \in \mathbb{L}^{\gamma^-}$.*

Proof. Let

$$G^{\alpha_k}(i,y) = c_i y + \alpha_k F(w^{\alpha_k})(i,y), \tag{6.31}$$

$$G(i,y) = c_i y + F(w^*)(i,y). \tag{6.32}$$

Because $w^*(i,\cdot)$ is K-convex, we know that a minimizer in (6.22) is given by

$$u^*(i,x) = \begin{cases} S_i - x, & x \le s_i, \\ 0, & x > s_i, \end{cases}$$

where S_i is the point where $G(i,y)$ defined by (6.32) attains its minimum, and s_i is a solution of the equation $G(i,y) = K + G(i,S_i)$ in y, if one exists, or $s_i = -\infty$ otherwise.

To prove that the strategy U^* is stable with respect to any $w \in \mathbb{L}^{\gamma-}$, we derive bounds for S_i and s_i. Obviously, S_i as a global minimizer of $G(i, y)$ is less than infinity for all i because $G(i, y) \to \infty$ for $y \to \infty$.

For $\alpha_0 \le \alpha < 1$, the (s_i^α, S_i^α) strategy is optimal, $v^\alpha(i, \cdot)$ is K-convex, and from Lemma 6.3 there is a $j \in I$ such that $s_j^\alpha \ge C_2$. We denote the smallest such j by $j(\alpha)$. For the sequence $(\alpha_k)_{k=1}^\infty$ from Theorem 6.1, at least one j^* appears infinitely often among the $j(\alpha_k)$. We choose a subsequence (still denoted by $(\alpha_k)_{k=1}^\infty$) such that $j(\alpha_k) = j^*$. Then, we know that for all $1 > \alpha_k \ge \alpha_0$, it holds that $s_{j^*}^\alpha \ge C_2$ and that $v^{\alpha_k}(j^*, x)$ is decreasing for $x \le s_{j^*}$ due to the K-convexity of $v^{\alpha_k}(j^*, x)$; (see proof of Theorem 3.4 in Chapter 3). Furthermore, by Lemma 6.5 we know that $S_{j^*}^\alpha \le C_8$ for $\alpha_2 \le \alpha < 1$.

Let $C_7 := \max\{S_{j^*}, C_8\}$. By the definition of s_i^α, we know that the argmin of $v^\alpha(i, x)$ is greater then or equal to s_i^α, and therefore is greater than or equal to C_2. Thus, we can conclude that for $1 > \alpha_k > \alpha_0$, we have

$$c_{j^*} C_2 + \alpha_k F(v^{\alpha_k})(j^*, C_2) - \min_{C_2 \le y \le C_7} \{c_{j^*} y + \alpha_k F(v^{\alpha_k})(j^*, y)\} \ge K.$$

Subtracting $\alpha_k v^{\alpha_k}(1, 0)$ from both the terms on the LHS, we obtain

$$G^{\alpha_k}(j^*, C_2) - \min_{C_2 \le y \le C_7} \{G^{\alpha_k}(j^*, y)\} \ge K,$$

where $G^{\alpha_k}(\cdot, \cdot)$ is defined by (6.31). The functions $G^{\alpha_k}(j^*, \cdot)$ converge uniformly to $G(j^*, \cdot)$ on any finite interval. Therefore, we can pass to a limit and get

$$G(j^*, C_2) - \min_{C_2 \le y \le C_7} \{G(j^*, y)\} \ge K. \tag{6.33}$$

Note that the minimum in (6.33) is attained at S_{j^*}. From (6.33) and the continuity of $G(j^*, \cdot)$, it follows that we can choose an $s_{j^*} \in [C_2, S_{j^*}]$ such that

$$G(j^*, s_{j^*}) - G(j^*, S_{j^*}) = K.$$

This means that under U^* there will be infinitely many nonzero orders. Because of the structure of the strategy, the inventory level under U^* will never be higher than $\max\{x, S_1, \ldots, S_L\}$. Therefore, we have

$$x_k \le \max\{x, S_1, \ldots, S_L\} < \infty, \quad \text{a.s.} \tag{6.34}$$

Furthermore, it follows from $s_{j^*} \ge C_2$ that if $i_n = j^*$, then $x_{n+1} \ge C_2 - \xi_n$. It follows from Assumption (viii) that the cumulative demand between two consecutive periods with demand state j^* has a finite γ^{th}

moment. This implies that $\mathsf{E}(x_k^\gamma)$ is uniformly bounded for all k. This, together with (6.34), proves stability of U^* with respect to any $w \in \mathbb{L}^{\gamma^-}$. □

LEMMA 6.10 *Let λ^* be defined as in Theorem 6.1. Then, for any admissible strategy U, $\lambda^* \leq J(i, x; U)$.*

Proof. Let $U = (u_0, u_1, \ldots)$ denote any admissible decision. Suppose

$$J(i, x; U) < \lambda^*. \tag{6.35}$$

Put $\tilde{f}(k) = \mathsf{E}[f(i_k, x_k) + c(i_k, u_k)]$. From (6.35), it immediately follows that $\sum_{k=0}^{n-1} \tilde{f}(k) < \infty$ for each positive integer n, since otherwise we would have $J(i, x; U) = \infty$. Note that

$$J(i, x; U) = \limsup_{n \to \infty} \frac{1}{n} \sum_{k=0}^{n-1} \tilde{f}(k),$$

while

$$(1 - \alpha) J^\alpha(i, x; U) = (1 - \alpha) \sum_{k=0}^{\infty} \alpha^k \tilde{f}(k). \tag{6.36}$$

Since $\tilde{f}(k)$ is nonnegative for each k, the sum in (6.36) is well-defined for $0 \leq \alpha < 1$, and we can use the Tauberian Theorem A.5.2 to obtain

$$\limsup_{\alpha \uparrow 1} (1 - \alpha) J^\alpha(i, x; U) \leq J(i, x; U) < \lambda^*.$$

On the other hand, we know from Theorem 3.2 that $(1-\alpha_k) v^{\alpha_k}(i, x) \to \lambda^*$ on a subsequence $\{\alpha_k\}_{k=1}^\infty$ converging to one. Thus, there exists an $\alpha < 1$ such that

$$(1 - \alpha) J^\alpha(i, x; U) < (1 - \alpha) v^\alpha(i, x),$$

which contradicts the definition of the value function $v^\alpha(i, x)$. □

THEOREM 6.3 (Verification Theorem)

(a) *Let $(\lambda, w(\cdot, \cdot))$ be a solution of the average cost optimality equation (6.22) with $w \in \mathbb{L}^{\gamma^-}$. Then, $\lambda \leq J(i, x; U)$ for any admissible U.*

(b) *Suppose there exists an $\hat{u}(i, x)$ for which the infimum in (6.22) is attained. Furthermore, let $\hat{U} = (\hat{u}, \hat{u}, \ldots)$, the stationary feedback policy given by $\hat{u}$, be stable with respect to w. Then,*

$$\lambda = J(i, x; \hat{U}) = \lambda^* = \lim_{N \to \infty} \frac{1}{N} \mathsf{E} \left(\sum_{k=0}^{N-1} f(i_k, \hat{x}_k) + c(i_k, \hat{u}_k) \right),$$

and $\hat{U}$ is an average optimal strategy.

(c) *Moreover, $\hat{U}$ minimizes*

$$\liminf_{N\to\infty} \frac{1}{N} \mathsf{E}\left(\sum_{k=0}^{N-1} f(i_k, x_k) + c(i_k, u_k)\right)$$

over the class of admissible decisions which are stable with respect to w.

Proof. We start by showing that

$$\lambda \le J(i, x; U) \text{ for any } U \text{ stable with respect to } w. \tag{6.37}$$

We assume that U is stable with respect to w, and then follow the same approach used in (2.15)) to obtain

$$\mathsf{E}\{w(i_{k+1}, x_{k+1}) \,|\, i_0, \ldots, i_k, \xi_0, \ldots, \xi_{k-1}\} = F(w)(i_k, x_k + u_k) \quad \text{a.s.} \tag{6.38}$$

Because u_k does not necessarily attain the infimum in (6.22), we have

$$w(i_k, x_k) + \lambda \le f(i_k, x_k) + c(i_k, u_k) + F(w)(i_k, x_k + u_k) \quad \text{a.s.},$$

and from (6.38) we derive

$$w(i_k, x_k) + \lambda \le f(i_k, x_k) + c(i_k, u_k) + \mathsf{E}(w(i_{k+1}, x_k + u_k - \xi_k) \,|\, i_k) \text{ a.s.}$$

By taking the expectation of both sides, we obtain

$$\mathsf{E}(w(i_k, x_k)) + \lambda \le \mathsf{E}(f(i_k, x_k) + c(i_k, u_k)) + \mathsf{E}(w(i_{k+1}, x_{k+1})).$$

Summing from 0 to $n-1$ yields

$$n\lambda \le \mathsf{E}\left(\sum_{k=0}^{n-1} f(i_k, x_k) + c(i_k, u_k)\right) + \mathsf{E}(w(i_n, x_n)) - \mathsf{E}(w(i_0, x_0)). \tag{6.39}$$

Divide by n, let n go to infinity, and use the fact that U is stable with respect to w to obtain

$$\lambda \le \liminf_{n\to\infty} \frac{1}{n} \mathsf{E}\left(\sum_{k=0}^{n-1} f(i_k, x_k) + c(i_k, u_k)\right). \tag{6.40}$$

Note that if the above inequality holds for 'liminf,' it certainly also holds for 'limsup'. This proves (6.37).

On the other hand, if there exists a $\hat{u}$ which attains the infimum in (6.22), we then have

$$w(i_k, \hat{x}_k) + \lambda = f(i_k, \hat{x}_k) + c(i_k, \hat{u}(i_k, \hat{x}_k)) + F(w)(i_k, \hat{x}_k + \hat{u}(i_k, \hat{x}_k)) \text{ a.s.},$$

and in a similar way we get

$$n\lambda = \mathsf{E}\left(\sum_{k=0}^{n-1} f(i_k, \hat{x}_k) + c(i_k, \hat{u}(i_k, \hat{x}_k))\right) \\ +\mathsf{E}(w(i_n, \hat{x}_n)) - \mathsf{E}(w(i_0, \hat{x}_0)). \tag{6.41}$$

Because $\hat{U}$ is assumed to be stable with respect to w, we get

$$\lambda = \lim_{n\to\infty} \frac{1}{n}\mathsf{E}\left(\sum_{k=0}^{n-1} f(i_k, \hat{x}_k) + c(i_k, \hat{u}(i_k, \hat{x}_k)\right) = J(i, x; \hat{U}). \tag{6.42}$$

For the special solution (λ^*, w^*) defined in Theorem 6.1 (see also Theorem 6.2), and the strategy U^* defined in Lemma 6.9, we get

$$\lambda^* = J(i, x; U^*).$$

Since by Lemma 6.9, U^* is stable with respect to any function in $\mathbb{L}^{\gamma^-}$, it follows that

$$\lambda \leq J(i, x; U^*) = \lambda^*, \tag{6.43}$$

which, in view of Lemma 6.10, proves part (a) of the theorem.

Part (a) of the theorem, together with (6.42), proves the average optimality of $\hat{U}$ over all admissible strategies. Furthermore, since $\lambda = J(i, x; \hat{U}) \geq \lambda^*$ by (6.42) and Lemma 6.10, it follows from (6.43) that $\lambda = \lambda^*$, and the proof of Part (b) is completed.

Finally, Part (c) immediately follows from Part (a) and (6.40). □

REMARK 6.4 It should be obvious that any solution (λ, w) of the average cost optimality equation and control $\hat{u}$ satisfying (a) and (b) of Theorem 6.3 will have a unique λ, since it represents the minimum average cost. On the other hand, if (λ, w) is a solution, then $(\lambda, w + c)$ is also a solution, where c is any constant. For the purpose of this chapter we do not require w to be unique up to a constant. If w is not unique up to a constant, then $\hat{u}$ may not be unique. We also do not need w^* in Theorem 6.1 to be unique.

The final result of this section, namely, that there exists an average optimal policy of (s, S)-type, is an immediate consequence of Lemmas 6.9 and 6.3.

THEOREM 6.4 *Let Assumptions* (i)-(viii) *hold. Let s_i and S_i, $i \in I$ be defined as in Lemma* 6.9. *Then, the stationary feedback strategy $U^* = (u^*, u^*, \ldots)$ defined by*

$$u^*(i, x) = \begin{cases} S_i - x, & x \leq s_i, \\ 0, & x > s_i \end{cases}$$

is average optimal.

The derivation of an optimal strategy presented above only uses properties of the function w^*. We want to mention here an additional way to obtain an average optimal state-dependent (s, S) policy.

COROLLARY 6.1 *Let $(\alpha_k)_{k=0}^{\infty}$ be a sequence of discount factors that converges to one and let (s_i, S_i) be any cluster point of the pairs of strategy parameters $(s_i^{\alpha_k}, S_i^{\alpha_k})$ on $[-\infty, \infty)$. Then, (s_i, S_i) is an average optimal policy.*

Proof. Because S_i^{α} is uniformly bounded from above for α sufficiently close to one, as proved in Lemma 6.5, a cluster point exists, if we allow the possibility of it being equal to $-\infty$. We choose a subsequence of the sequence $(\alpha_k)_{k=0}^{\infty}$ defined in Theorem 6.1, still denoted by $(\alpha_k)_{k=0}^{\infty}$, such that the strategy parameters $s_i^{\alpha_k}$ and $S_i^{\alpha_k}$ converge to s_i and S_i, and w^{α_k} converges to w^*. Then, by Theorem 6.2, w^* satisfies the average optimality equation (6.22). Furthermore, it is easy to show that for any fixed $x \neq s_i$, $u^{\alpha_k}(i, x)$ converges to

$$u^*(i, x) = \begin{cases} 0, & x > s_i \\ S_i - x, & x \leq s_i. \end{cases}$$

For $x = s_i$, we can choose at least one of the following types of subsequences:

(a) $s_i^{\alpha_k} < x$ for all k, or

(b) $s_i^{\alpha_k} \geq x$ for all k.

For the type (a) subsequence we obtain $u^{\alpha_k}(i, x) \to x$, and for type (b) we get $u^{\alpha_k}(i, x) \to S_i - x$. By Theorem 6.2, the limit $u^*(i, x)$ is precisely the function for which the infimum in (6.22) is attained. Then, by Theorem 6.3, the policy $U^* = (u^*(i_0, x_0), u^*(i_1, x_1), \ldots)$ is average optimal. This proves the optimality of at least one of the two types of (s, S) strategies defined in (3.46) and (3.50), respectively. On the other hand, because of the continuity of w^*, if one of the types is optimal, then so is the other type with the same parameters (s_i, S_i). □

6.5. Concluding Remarks and Notes

In this chapter based on Beyer *et al.* (1998), we have generalized the average cost infinite horizon inventory models involving fixed cost that have appeared in the literature to allow for unbounded demands and costs with polynomial growth. We have shown the existence of an optimal Markov policy, and that this can be a state-dependent (s, S) policy.

In the vanishing discount approach to the average cost problem, relaxation of the bounded Markovian demand assumption in Chapter 5 to unbounded Markovian demands requires the use of renewal arguments for the establishment of the crucial local equi-Lipschitzian property of the discounted cost value function. This has been carried out in Lemmas 6.1–6.4. Finally, the establishment of the ergodic dynamic programming equation requires proving uniform integrability of the differential discounted value functions, accomplished in Lemmas 6.6, 6.7 and Theorem 6.2, and the uniform boundedness of the discounted order-up-to-levels accomplished in Lemma 6.5.

The question of whether or not the potential function w^* is unique up to a constant remains open. The considerable literature on average cost MDPs, surveyed in Arapostathis *et al.* (1993) and Hernández-Lerma and Lasserre (1996), could be found useful in addressing the question.

Chapter 7

AVERAGE COST MODELS WITH LOST SALES

7.1. Introduction

This chapter is concerned with the long-run average cost minimization of a stochastic inventory problem with Markovian demand, fixed ordering cost, and convex surplus cost in the case of lost sales. The formulation of the problem is similar to that introduced in Chapter 4 except that we replace the discounted cost objective function by the long-run average cost objective function. To deal with this average cost problem, we apply the vanishing discount method to solve the dynamic programming equations defined for the problem, and establish the corresponding verification theorem.

The plan of this chapter is as follows. Required results for the discounted cost model derived in Chapter 4 are recapitulated in Section 7.3. In the next section, we provide a precise formulation of the problem. In Section 7.4, we obtain the asymptotic behavior of the differential discounted value function as the discount rate goes to zero. The vanishing discount approach to establish the average cost optimality equation is developed in Section 7.5. The associated verification theorem is proved in Section 7.6, and the theorem is used to show that a state-dependent (s, S) policy is optimal for the problem. Section 7.7 concludes the chapter with suggestions for future research.

7.2. Formulation of the Model

Consider an inventory problem over an infinite horizon. The demand in each period is assumed to be a random variable defined on a given probability space, and not necessarily identically distributed. To precisely define the demand process, we consider a finite collection

D. Beyer et al., *Markovian Demand Inventory Models*, International Series in Operations Research & Management Science 108,
DOI 10.1007/978-0-387-71604-6_7,

of demand states labeled $i \in \mathbb{I} = \{1, 2, \ldots, L\}$, and let i_k denote the demand state observed at the beginning of period k. We assume that i_k, $k = 0, 1, 2, \ldots$, is a Markov chain over $\mathbb{I}$, with the transition matrix $P = \{p_{ij}\}$. Thus,

$$0 \leq p_{ij} \leq 1, \; i \in \mathbb{I}, j \in \mathbb{I}, \; \text{ and } \sum_{j=1}^{L} p_{ij} = 1, \; i \in \mathbb{I}.$$

Let the nonnegative random variable ξ_k denote the demand at the end of a given period $k \in \langle 0, N-1 \rangle$. Demand ξ_k depends only on the demand state in period k, by which we mean that it does does not depend on k and is independent of past demand states and past demands. We denote its probability density by $\varphi_i(x)$ and its probability distribution by $\Phi_i(x)$ when the demand state $i_k = i$.

We suppose that orders are placed at the beginning of a period, delivered instantaneously, and followed by the period's demand. Unsatisfied demands are lost.

In what follows, we list the assumptions that are needed to derive the main results of the chapter in Sections 7.5 and 7.6. Because not all the results proved in this chapter require all of these assumptions, we label them as follows so that we can specify the assumptions required in the statements of the specific results proved in this chapter.

(i) The production cost is given by $c(i, u) = K\mathbb{I}_{u>0} + c_i u$, where $K \geq 0$ is the fixed ordering cost and $c_i \geq 0$ is the variable cost.

(ii) For each i, the inventory cost function $f(i, \cdot)$ is convex, nondecreasing and of linear growth, i.e., $f(i, x) \leq C_f(1 + |x|)$ for some $C_f > 0$ and all x. Also, $f(i, x) = 0$ for all $x \leq 0$.

(iii) For each i, the shortage cost function $q(i, \cdot)$ is convex, nonincreasing and of linear growth, i.e., $q(i, x) \leq C_q(1 + |x|)$ for some $C_q > 0$ and all x. Also, $q(i, x) = 0$ for all $x \geq 0$.

(iv) There is a state $g \in \mathbb{I}$ such that $f(g, x)$ is not identically zero.

(v) The production and inventory costs satisfy for all i,

$$c_i x + \sum_{j=1}^{L} p_{ij} \int_0^{\infty} f(j, (x - z)^+) d\Phi_i(z) \to \infty \text{ as } x \to \infty. \qquad (7.1)$$

(vi) The Markov chain $(i_k)_{k=0}^{\infty}$ is irreducible.

(vii) There is a state $h \in \mathbb{I}$ such that $1 - \Phi_h(\varepsilon) = \rho > 0$ for some $\varepsilon > 0$.

(viii) For each i, the inequality $q'^{-}(i,0) \leq \bar{f}'^{+}(i,0) - \bar{c}^i$ holds.

(ix) $\mathsf{E}\, \xi_k \leq D < \infty$ for all k.

REMARK 7.1 Assumptions (i)–(iii) reflect the usual structure of the production and inventory costs to prove the optimality of an (s_i, S_i) policy. Note that K is the same for all i. In the stationary case, this is equivalent to the condition (2.18) required in the nonstationary model for the existence of an optimal (s_i, S_i) policy; (see Chapter 2). Assumption (iv) rules out trivial cases where the optimal policy is never to order. Assumption (v) means that either the unit ordering cost $c_i > 0$ or the second term in (7.1), which is the expected holding cost, or both, go to infinity as the surplus level x goes to infinity. While related, Assumption (v) neither implies nor is implied by Assumption (iv). Assumption (v) is borne out of practical considerations and is not very restrictive. In addition, it rules out such unrealistic trivial cases as the one with $c_i = 0$ and $f(i,x) = 0, x \geq 0$, for each i, which implies ordering an infinite amount whenever an order is placed. Assumptions (iv) and (v) generalize the usual assumption made by Scarf (1960) and others that the unit inventory holding cost $h > 0$. Assumption (viii) replaces the more stringent assumption in the classical literature that the unit shortage cost is not smaller than the unit price.

REMARK 7.2 Assumptions (vi) and (vii) are needed to deplete any given initial inventory in a finite expected time. While Assumption (vii) says that in at least one state h, the expected demand is strictly larger than zero, Assumption (vi) implies that the state h would occur infinitely often with finite expected intervals between successive occurrences.

REMARK 7.3 Assumption (viii) means that the marginal shortage cost in one period is larger than or equal to the expected unit ordering cost less the expected marginal inventory holding cost in any state of the next period. If this condition does not hold, that is, if $-q_n'^{-}(i,0) < \bar{c}^i_{n+1} - \bar{f}'^{+}_{n+1}(i,0)$ for some i, a speculative retailer may find it attractive to meet a smaller part of the demand in period n than is possible from the available stock, carry the leftover inventories to period $(n+1)$, and order a little less as a result in period $(n+1)$, with the expectation that he will be better off. Thus, Assumption (viii) rules out this kind of speculation on the part of the retailer. But such a speculative behavior is not allowed in our formulation of the dynamics in any case, since the demand in any period must be satisfied to the extent of the availability of inventories. This suggests that it might be possible to prove our results without Assumption (viii).

The objective is to minimize the expected long-run average cost

$$J(i,x;U) = \limsup_{N\to\infty} \frac{1}{N} \mathsf{E}\Big\{ \sum_{k=0}^{N-1} [c(i_k,u_k) + f(i_k,x_k) + q(i_k, x_k+u_k-\xi_k)] \Big\}, \tag{7.2}$$

with $i_0 = i$ and $x_0 = x \geq 0$, where $U = (u_0, u_1, \ldots)$, $u_i \geq 0$, $i = 0, 1, \ldots,$ is a *history-dependent* or *nonanticipative* decision (order quantities) for the problem. Such a control U is termed admissible. Let $\mathcal{U}$ denote the class of all admissible controls. The surplus balance equations are given by

$$x_{k+1} = (x_k + u_k - \xi_k)^+, \; k = 0, 1, \ldots. \tag{7.3}$$

Our aim is to show that there exist a constant λ^*, termed the optimal average cost, which is independent of the initial i and x, and a control $U^* \in \mathcal{U}$ such that

$$\lambda^* = J(i,x;U^*) \leq J(i,x;U) \quad \text{for all} \quad U \in \mathcal{U}, \tag{7.4}$$

and

$$\lambda^* = \lim_{N\to\infty} \frac{1}{N} \mathsf{E}\Big\{ \sum_{k=0}^{N-1} [c(i_k,u_k^*) + f(i_k,x_k^*) + q(i_k, x_k+u_k-\xi_k)] \Big\}, \tag{7.5}$$

where x_k^*, $k = 0, 1, \ldots$, is the surplus process corresponding to U^* with $i_0 = i$ and $x_0 = x$.

To prove these results, we will use the vanishing discount approach. That is, by letting the discount factor α in the discounted cost problem approach one, we will show that we can derive a dynamic programming equation whose solution provides an average optimal control and the associated minimum average cost λ^*.

For this purpose, we recapitulate relevant results for the discounted cost problem obtained in Chapter 4.

REMARK 7.4 Note that the objective function (7.2) is slightly, but not essentially, different from that used in the classical literature. Whereas we base the surplus cost on the initial surplus in each period, the usual practice in the literature is to charge the cost on the ending surplus levels, which means to have $f(i_k, x_{k+1})$ instead of $f(i_k, x_k)$ in (7.2). Note that

x_{k+1} is also the ending inventory in period k. It should be obvious that this difference in the objective functions does not change the long-run average cost for any admissible policy. By the same token, we can justify our choice to charge shortage costs at the end of a given period.

7.3. Discounted Cost Model Results from Chapter 4

Consider the model formulated above with the average cost objective (7.2) replaced by the extended real-valued objective function

$$J^\alpha(i,x;U) = \sum_{k=0}^{\infty} \alpha^k \mathsf{E}[c(i_k,u_k)+f(i_k,x_k)+q(i_k,x_k+u_k-\xi_k)], \quad 0 \le \alpha < 1. \tag{7.6}$$

Define the value function with $i_0 = i$ and $x_0 = x$ as

$$v^\alpha(i,x) = \inf_{U \in \mathcal{U}} J^\alpha(i,x;U). \tag{7.7}$$

Let $\mathbb{B}_0$ denote the class of all continuous functions from $I \times \mathbb{R}$ into $[0,\infty)$, and the pointwise limits of sequences of these functions; (see Feller (1971)). Note that it includes piecewise-continuous functions. Let $\mathbb{B}_1$ denote the space of functions in $\mathbb{B}_0$ that are of linear growth, i.e., for any $b \in \mathbb{B}_1$, $0 \le b(i,x) \le C_b(1+|x|)$ for some $C_b > 0$. Let $\mathbb{B}_2$ denote the subspace of functions in $\mathbb{B}_1$ that are uniformly continuous with respect to $x \in R$. For any $b \in \mathbb{B}_1$, we define

$$F(b)(i,y) = \sum_{j=1}^{L} p_{ij} \int_0^M b(j,(y-z)^+)d\Phi_i(z). \tag{7.8}$$

THEOREM 7.1 *Let Assumptions* (i)–(iii), (v), (viii), *and* (ix) *hold. Then, we have the following results.*

(a) *The value function $v^\alpha(\cdot,\cdot)$ is in $\mathbb{B}_2$, and it solves the dynamic programming equation*

$$\begin{aligned} v^\alpha(i,x) &= f(i,x) + \inf_{u \ge 0}\{c(i,u) + \mathsf{E}[q(i,x+u-\xi^i) \\ &\quad + \alpha \sum_{j=1}^{L} p_{ij} v^\alpha(j,(x+u-\xi^i)^+)]\} \\ &= f(i,x) + \inf_{u\ge 0}\{c(i,u) + \mathsf{E}q(i,x+u-\xi^i) \\ &\quad + \alpha F(v^\alpha)(j,x+u)\}. \end{aligned} \tag{7.9}$$

(b) $v^\alpha(i,\cdot)$ *is* K*-convex and there are real numbers* (s_i^α, S_i^α), $s_i^\alpha \le S_i^\alpha$, *such that the feedback policy* $\hat{u}_k^\alpha(i,x) = (S_i^\alpha - x)\mathbb{1}_{x<s_i^\alpha}$ *is optimal.*

Proof. (Actually Theorem 7.1 has been stated but not proved in Chapter 4.) The proof of part (a) follows the lines of the proof of Theorem 3.3 in Chapter 3 by taking the limit of the n-period value function, for n tending to infinity. Part (b) immediately follows since the limit of a sequence of K-convex functions is K-convex. □

7.4. Limiting Behavior as the Discount Factor Approaches 1

Hereafter, we will omit the additional superscript α on the control policies for ease of notation. Thus, for example, $\hat{u}_k^\alpha(i,x)$ will be denoted simply as $\hat{u}_k(i,x)$. Since we do not consider the limits of the control variables as $\alpha \to 1$, the practice of omitting the superscript α will not cause any confusion. In any case, the dependence of controls on α will always be clear from the context.

To insure a "smooth" limiting behavior for $\alpha \to 1$, we prove in Lemma 7.2 that $v^\alpha(i,\cdot)$ is locally equi-Lipschitzian. For this we need some notation and a preliminary result. For any $y > 0$, let

$$\tau_y := \inf\{n : \sum_{k=0}^{n} \xi_k \ge y\}$$

be the first index for which the cumulative demand is not less than y. The following required result is proved in Chapter 5.

LEMMA 7.1 *Let Assumptions* (vii) *and* (viii) *hold. Then, for any* $l \in \mathbb{I}$, *we have* $\mathsf{E}(\tau_y | i_0 = l) < \infty$.

LEMMA 7.2 *Under Assumptions* (i)-(iii), (vi) , (vii), *and* (ix), $v^\alpha(i,\cdot)$ *is locally equi-Lipschitzian, i.e., for* $X > 0$ *there is a positive constant* $C^1 < \infty$, *independent of* α, *such that*

$$|v^\alpha(i,x) - v^\alpha(i,\tilde{x})| \le C^1 |x - \tilde{x}| \quad \textit{for all} \quad x, \tilde{x} \in [0, X]. \tag{7.10}$$

Proof. Consider the case $\tilde{x} \ge x$. Let us fix an $\alpha \in [0,1)$. It follows from Theorem 7.1 that there is an optimal feedback strategy U. Use the strategy U with initial surplus x, and the strategy $\tilde{U}$ defined by

$$\tilde{u}_k = [u_k - (\tilde{x}_k - x_k)]^+ = \begin{cases} 0 & \text{if } \; u_k \le \tilde{x}_k - x_k, \\ u_k + x_k - \tilde{x}_k & \text{if } \; u_k > \tilde{x}_k - x_k, \end{cases}$$

with initial $\tilde{x}$, x_k, and $\tilde{x}_k$ denoting the inventory levels resulting from the respective strategies. It is easy to see that the inequalities

$$0 \le \tilde{x}_k - x_k \le \tilde{x} - x \quad \text{and} \quad \tilde{u}_k \le u_k$$

hold for all k. Let $\tilde{\tau} := \tau_{\tilde{x}}$ for ease of notation. If $\tilde{u}_k = 0$ for all $k \in [0, \tilde{\tau}]$, then $\tilde{x}_{\tilde{\tau}} = x_{\tilde{\tau}} = 0$, and the two trajectories are identical for all $k > \tilde{\tau}$. If $\tilde{u}_{k'} \ne 0$ for some $k' \in [0, \tilde{\tau}]$, then $\tilde{x}_{k'} = x_{k'}$, and the two trajectories are identical for all $k > k'$. In any case, the two trajectories are identical for all $k > \tilde{\tau}$. From Assumptions (i)–(iii), we have

$$c(i_k, \tilde{u}_k) \le c(i_k, u_k),$$

$$|f(i_k, \tilde{x}_k) - f(i_k, x_k)| \le C_f |\tilde{x} - x|,$$

$$|q(i_k, \tilde{x}_k + \tilde{u}_k - \xi_k) - q(i_k, x_k + u_k - \xi_k)| \le C_q |\tilde{x} - x|.$$

Therefore,

$$\begin{aligned} & v^\alpha(i, \tilde{x}) - v^\alpha(i, x) \\ \le\ & J^\alpha(i, \tilde{x}; \tilde{U}) - J^\alpha(i, x; U) \\ =\ & \mathsf{E}\Big(\sum_{k=0}^{\tilde{\tau}} \alpha^k (f(i_k, \tilde{x}_k) - f(i_k, x_k) + q(i_k, \tilde{x}_k + \tilde{u}_k - \xi_k) \\ & -q(i_k, x_k + u_k - \xi_k) + c(i_k, \tilde{u}_k) - c(i_k, u_k))\Big) \\ \le\ & \mathsf{E}\Big(\sum_{k=0}^{\tilde{\tau}} \alpha^k (C_f + C_q)|\tilde{x} - x|\Big) \\ \le\ & \mathsf{E}(\tilde{\tau} + 1)(C_f + C_q)|\tilde{x} - x|. \end{aligned} \tag{7.11}$$

It immediately follows from Lemma 7.1 that $\mathsf{E}(\tilde{\tau}+1) = \mathsf{E}(\tau_{\tilde{x}} + 1 | i_0 = i) \le \mathsf{E}(\tau_x + 1 | i_0 = i) < \infty$.

To complete the proof, it is sufficient to prove the above inequality for $\tilde{x} < x$. In this case, let us define the strategy $\tilde{U}$ by

$$\tilde{u}_k = \begin{cases} u_k + x - \tilde{x} & \text{if } u_k > 0, \\ 0 & \text{otherwise,} \end{cases}$$

It is easy to see that the inequalities

$$0 \ge \tilde{x}_k - x_k \ge \tilde{x} - x \quad \text{and} \quad \tilde{u}_k - u_k \le x - \tilde{x}$$

hold for all k. Let $\tau := \tau_x$ for ease of notation. If $u_k = 0$ for all $k \in [0, \tau]$, then $\tilde{x}_\tau = x_\tau = 0$, and the two trajectories are identical for all $k > \tau$. If $u_{k'} \ne 0$ for some $k' \in [0, \tau]$, then $\tilde{x}_{k'} = x_{k'}$, and the two trajectories are identical for all $k > k'$. In either case, the two trajectories are identical for all $k > \tau$.

From Assumptions (i)–(iii), we have

$$c(i_k, \tilde{u}_k) - c(i_k, u_k) \le \max\{c_i\}|x - \tilde{x}|,$$

$$|f(i_k, \tilde{x}_k) - f(i_k, x_k)| \le C_f|\tilde{x} - x|,$$

$$|q(i_k, \tilde{x}_k + \tilde{u}_k - \xi_k) - q(i_k, x_k + u_k - \xi_k)| \le C_q|\tilde{x} - x|.$$

Therefore,

$$\begin{aligned} & v^\alpha(i, \tilde{x}) - v^\alpha(i, x) \\ \le\ & J^\alpha(i, \tilde{x}; \tilde{U}) - J^\alpha(i, x; U) \\ =\ & \mathsf{E}\Big(\sum_{k=0}^{\tau} \alpha^k (f(i_k, \tilde{x}_k) - f(i_k, x_k) + q(i_k, \tilde{x}_k + \tilde{u}_k - \xi_k) \\ & -q(i_k, x_k + u_k - \xi_k) + c(i_k, \tilde{u}_k) - c(i_k, u_k))\Big) \\ \le\ & \mathsf{E}\Big(\sum_{k=0}^{\tau} \alpha^k (C_f + C_q + \max\{c_i\})|\tilde{x} - x|\Big) \\ \le\ & \mathsf{E}(\tau + 1)(C_f + C_q + \max\{c_i\})|\tilde{x} - x|. \end{aligned} \tag{7.12}$$

Lemma 7.1 implies again that $\mathsf{E}(\tau+1) = \mathsf{E}(\tau_x+1|i_0 = i) \le \mathsf{E}(\tau_{\tilde{x}}+1|i_0 = i) < \infty$ and the proof is complete. □

LEMMA 7.3 *Under Assumptions* (i)–(ix), *there are constants* $\alpha_0 \in [0, 1)$ *and* $C_2 > 0$ *such that for all* $\alpha \ge \alpha_0$ *and for any* i *for which* $s_i^\alpha > 0$, *we have* $S_i^\alpha \le C_2 < \infty$.

Proof. Let us fix the initial state $i_0 = i$ for which $s_i^\alpha > 0$. Fix $\alpha_0 > 0$ and a discount factor $\alpha \ge \alpha_0$. Let $U = (u(i_0, x_0), u(i_1, x_1), \ldots)$ be an optimal strategy with parameters (s_j^α, S_j^α), $j \in I$. Let us fix a positive real number Y and assume $S_i^\alpha > Y$. In what follows, we specify a value of Y, namely Y^*, in terms of which we will construct an alternative strategy $\tilde{U}$ that is better than U.

For the demand state g specified in Assumption (iv), let

$$\tau^g := \inf\{n > 0 : i_n = g\}$$

be the first period (not counting the period 0) with the demand state g. Furthermore, let d be the state with the lowest per unit ordering cost, i.e., $c_d \le c_i$ for all $i \in I$. Then we define

$$\tau := \inf\{n \ge \tau^g : i_n = d\}.$$

Assume $x_0 = \tilde{x}_0 = \bar{x} := 0$, and consider the policy $\tilde{U}$ defined by

$$\begin{cases} \tilde{u}_k = 0, \; k = 0, 1, 2, \ldots, \tau - 1, \\ \tilde{u}_\tau = x_\tau + u(i_\tau, x_\tau), \\ \tilde{u}_k = u(i_k, x_k), \; k \geq \tau + 1. \end{cases}$$

The two policies and the resulting trajectories differ only in the periods 0 through τ. Therefore, we have

$$\begin{aligned} & v^\alpha(i, \bar{x}) - J^\alpha(i, \bar{x}; \tilde{U}) \\ &= J^\alpha(i, \bar{x}; U) - J^\alpha(i, \bar{x}; \tilde{U}) \\ &= \mathsf{E}\Big(\sum_{k=0}^{\tau} \alpha^k (f(i_k, x_k) - f(i_k, \tilde{x}_k) + q(i_k, x_k + u_k - \xi_k) \\ &\quad -q(i_k, \tilde{x}_k + \tilde{u}_k - \xi_k) + c(i_k, u_k) - c(i_k, \tilde{u}_k)) \Big) \\ &= \mathsf{E}\left(\sum_{k=1}^{\tau} \alpha^k (f(i_k, x_k) + q(i_k, x_k + u_k - \xi_k) - q(i_k, -\xi_k)) \right) \\ &\quad + \mathsf{E}\left(\sum_{k=0}^{\tau} \alpha^k c(i_k, u_k) \right) - \mathsf{E}(\alpha^\tau c(i_\tau, \tilde{u}_\tau)). \end{aligned} \tag{7.13}$$

After ordering in period τ, the total accumulated ordered amount up to period τ is less for the policy $\tilde{U}$ than it is for U. Observe that the policy $\tilde{U}$ orders only in the period τ or later. The order of the policy $\tilde{U}$ is executed at the lowest possible per unit cost c_d in the period τ, which is not earlier than any of the ordering periods of policy U. Because $\tilde{U}$ orders only once and U orders at least once in periods $0, 1, \ldots, \tau$, the total fixed ordering cost of $\tilde{U}$ does not exceed the total fixed ordering cost of U. Thus,

$$\mathsf{E}\left(\sum_{k=0}^{\tau} \alpha^k c(i_k, u_k) \right) \geq \mathsf{E}(\alpha^\tau c(i_\tau, \tilde{u}_\tau)).$$

Furthermore, it follows from Assumptions (iii), (vi), and (ix) that

$$\mathsf{E}\left(\sum_{k=1}^{\tau} q(i_k, -\xi_t) \right) < \infty.$$

Because $\tau \geq \tau^g$, we obtain

$$\mathsf{E}\left(\sum_{k=1}^{\tau} \alpha^k (f(i_k, x_k) + q(i_k, x_k + u_k - \xi_k)) \right) \geq \mathsf{E}\left(\sum_{k=1}^{\tau^g} \alpha^k f(i_k, x_k) \right),$$

and because $S_i^\alpha \geq Y$, we obtain

$$x_k \geq Y - \sum_{t=1}^{k} \xi_t.$$

Irreducibility of the Markov chain $(i_n)_{n=0}^{\infty}$ implies existence of an integer m, $0 \leq m \leq L$, such that $\mathsf{P}(i_m = g) > 0$. Let m_0 be the smallest such m. It follows that $\tau^g \geq m_0$ and therefore, for all $\alpha \geq \alpha_0$

$$\begin{aligned} &\mathsf{E}\left(\sum_{k=1}^{\tau^g} \alpha^k f(i_k, x_k)\right) \\ &\geq \alpha^{m_0} \mathsf{E}(f(i_{m_0}, x_{m_0})) \\ &\geq \alpha_0^{m_0} \mathsf{E}(f(g, Y - \sum_{t=1}^{m_0} \xi_t) | i_{m_0} = g) \mathsf{P}(i_{m_0} = g). \end{aligned} \tag{7.14}$$

Using Assumptions (ii), (iv), and (ix), it is easy to show that the RHS of (7.14) tends to infinity as Y goes to infinity. Therefore, we can choose Y^*, $0 \leq Y^* < \infty$ such that for all $\alpha \geq \alpha_0$,

$$\begin{aligned} v^\alpha(i, \bar{x}) - J^\alpha(i, \bar{x}; \tilde{U}) \geq\ & \alpha_0^{m_0} \mathsf{E}(f(g, Y^* - \sum_{t=1}^{m_0} \xi_t) | i_{m_0} = g) \mathsf{P}(i_{m_0} = g) \\ & - \mathsf{E}\left(\sum_{k=1}^{\tau} q(i_k, -\xi_t)\right) > 0. \end{aligned} \tag{7.15}$$

Note that the RHS of (7.15) is independent of α. Therefore, for $\alpha \geq \alpha_0$, a policy with $S_i^\alpha > C_2 := Y^*$ cannot be optimal. □

7.5. Vanishing Discount Approach

LEMMA 7.4 *Under Assumptions* (i)-(ix), *the differential discounted value function* $w^\alpha(i, x) := v^\alpha(i, x) - v^\alpha(1, 0)$ *is uniformly bounded with respect to* α *for all* x *and* i.

Proof. Since Lemma 7.2 implies

$$\begin{aligned} |w^\alpha(i, x)| &= |v^\alpha(i, x) - v^\alpha(1, 0)| \\ &\leq |v^\alpha(i, x) - v^\alpha(i, 0)| + |v^\alpha(i, 0) - v^\alpha(1, 0)| \\ &\leq C_3 |x| + |w^\alpha(i, 0)|, \end{aligned}$$

it is sufficient to prove that $w^\alpha(i, 0)$ is uniformly bounded. Note that C_3 may depend on x, but it is independent of α.

First, we show that there is an $\underline{M} > -\infty$ with $w^\alpha(i,0) \geq \underline{M}$ for all α. Let α be fixed. From Theorem 7.1 we know that in this discounted case there is a stationary optimal feedback policy $U = (u(i,x), u(i,x), \ldots)$. With $k^* = \inf\{k : i_k = i\}$, we consider the cost for the initial state $(i_0, \tilde{x}_0) = (1,0)$, and the inventory policy $\tilde{U}$ that does not order in periods $0, 1, \ldots, k^* - 1$, and follows U starting from the period k^*, i.e., $\tilde{U}$ is defined by

$$\begin{cases} \tilde{u}_k = 0 \text{ for } k < k^*, \\ \tilde{u}_k = u(i_k, x_k) \text{ for } k \geq k^*. \end{cases}$$

The cost corresponding to this policy is

$$J^\alpha(1,0;\tilde{U}) = \mathsf{E}\left(\sum_{k=0}^{k^*-1} \alpha^k q(i_k, -\xi_k) + \alpha^{k^*} v^\alpha(i,0)\right). \tag{7.16}$$

Because of Assumptions (iii), (vi) and (ix), there exists a constant $\underline{M}$ such that

$$\mathsf{E}\left(\sum_{k=0}^{k^*-1} q(i_k, -\xi_k)\right) \leq -\underline{M} < \infty.$$

Therefore, we have

$$\begin{aligned} w^\alpha(i,0) &= v^\alpha(i,0) - v^\alpha(1,0) \geq v^\alpha(i,0) - J^\alpha(1,0;\tilde{U}) \\ &\geq v^\alpha(i,0) - \mathsf{E}\left(\sum_{k=0}^{k^*-1} \alpha^k q(i_k, -\xi_k) + \alpha^{k^*} v^\alpha(i,0)\right) \\ &\geq v^\alpha(i,0)(1 - \mathsf{E}(\alpha^{k^*})) + \underline{M} \geq \underline{M}. \end{aligned} \tag{7.17}$$

The validity of the inequality $w^\alpha(i,0) \leq \overline{M}$ is shown analogously by changing the role of the states 1 and i. Thus,

$$|w^\alpha(i,x)| \leq C_3|x| + \max\{\underline{M}, \overline{M}\},$$

and the proof is complete. □

LEMMA 7.5 *Under Assumptions* (iii) *and* (viii), $(1-\alpha)v^\alpha(1,0)$ *is uniformly bounded for* $0 < \alpha < 1$.

Proof. Consider the strategy $\mathbf{0} = (0, 0, \ldots)$. Then, because $\mathbf{0}$ is not necessarily optimal,

$$0 \leq v^\alpha(1,0) \leq J^\alpha(1,0;\mathbf{0}) = \mathsf{E}\left(\sum_{k=0}^{\infty} \alpha^k q(i_k, -\xi_k)\right).$$

Because of Assumptions (iii) and (ix), $\mathsf{E}q(i,-\xi_k)$ is bounded for all i and there is a $C_4 < \infty$ such that $\mathsf{E}(q(i_k,-\xi_k)) < C_4$. Therefore,

$$0 \leq (1-\alpha)v^{\alpha}(1,0) \leq (1-\alpha)\sum_{k=0}^{\infty}\alpha^k C_4 = C_4.$$

□

THEOREM 7.2 *Let Assumptions* (i)–(ix) *hold. There exist a sequence* $(\alpha_k)_{k=1}^{\infty}$ *converging to* 1, *a constant* λ^*, *and a locally Lipschitz continuous function* $w^*(\cdot,\cdot)$, *such that*

$$(1-\alpha_k)v^{\alpha_k}(i,x) \to \lambda^* \quad \textit{and} \quad w^{\alpha_k}(i,x) \to w^*(i,x),$$

locally uniformly in x *and* i *as* k *goes to infinity. Moreover,* (λ^*, w^*) *satisfies the average cost optimality equation*

$$\begin{aligned} w(i,x)+\lambda &= f(i,x)+\inf_{u\geq 0}\{c(i,u)+\mathsf{E}q(i,x+u-\xi^i) \\ &\quad +F(w)(i,x+u)\}. \qquad (7.18)\end{aligned}$$

Proof. It is immediate from Lemma 7.2 and the definition of $w^{\alpha}(i,x)$ that $w^{\alpha}(i,\cdot)$ is locally equi-Lipschitzian for $\alpha \geq \alpha_0$, and therefore it is uniformly continuous on any finite interval. Additionally, according to Lemma 7.4, $w^{\alpha}(i,\cdot)$ is uniformly bounded, and by Lemma 7.5, $(1-\alpha)v^{\alpha}(1,0)$ is also uniformly bounded. Therefore, the Arzelà-Ascoli Theorem A.3.5 and Lemma 7.2, lead to the existence of a sequence $\alpha_k \to 1$, a locally Lipschitz continuous function $w^*(i,x)$, and a constant λ^* such that

$$(1-\alpha_k)v^{\alpha_k}(1,0) \to \lambda^* \quad \text{and} \quad w^{\alpha_k}(i,x) \to w^*(i,x),$$

with the convergence of $w^{\alpha_k}(i,x)$ to $w^*(i,x)$ being locally uniform.

It is easily seen that

$$\lim_{k\to\infty}(1-\alpha_k)v^{\alpha_k}(i,x) = \lim_{k\to\infty}(1-\alpha_k)(w^{\alpha_k}(i,x)+v^{\alpha_k}(1,0)) = \lambda^*.$$

Substituting $v^{\alpha_k}(i,x) = w^{\alpha_k}(i,x)+v^{\alpha_k}(1,0)$ into (7.9) yields

$$\begin{aligned} w^{\alpha_k}(i,x) \quad + \quad (1-\alpha_k)v^{\alpha_k}(1,0) &= f(i,x)+\inf_{u\geq 0}\{c(i,u) \\ &\quad +\mathsf{E}q(i,x+u-\xi^i)+\alpha_k F(w^{\alpha_k})(i,x+u)\}. \qquad (7.19)\end{aligned}$$

Since $w^{\alpha_k}(i,x)$ converges locally uniformly with respect to x and i and since for any given x a minimizer u^* in (7.19) can be chosen such that,

$x + u^* - \xi \in [0, x + C_2]$, we can use Lemma 7.3 and pass to a limit on both sides of (7.19) to obtain (7.18). This completes the proof. □

LEMMA 7.6 *Let λ^* be defined as in Theorem 7.2. Let Assumptions (i)–(ix) hold. Then, for any admissible strategy U, we have $\lambda^* \leq J(i, x; U)$.*

Proof. Let $U = (u_0, u_1, \ldots)$ denote any admissible decision. Suppose

$$J(i, x; U) < \lambda^*. \tag{7.20}$$

Put

$$\tilde{f}(k) = \mathsf{E}[f(i_k, x_k) + q(i_k, x_k + u_k - \xi_k) + c(i_k, u_k)].$$

From (7.20), it immediately follows that $\sum_{k=0}^{n-1} \tilde{f}(k) < \infty$ for each positive integer n, since otherwise we would have $J(i, x; U) = \infty$. Note that

$$J(i, x; U) = \limsup_{n \to \infty} \frac{1}{n} \sum_{k=0}^{n-1} \tilde{f}(k),$$

while

$$(1 - \alpha) J^\alpha(i, x; U) = (1 - \alpha) \sum_{k=0}^{\infty} \alpha^k \tilde{f}(k). \tag{7.21}$$

Since $\tilde{f}(k)$ is nonnegative for each k, the sum in (7.21) is well-defined for $0 \leq \alpha < 1$, and we can use the Tauberian Theorem A.5.2 to obtain

$$\limsup_{\alpha \uparrow 1} (1 - \alpha) J^\alpha(i, x; U) \leq J(i, x; U) < \lambda^*.$$

On the other hand, we know from Theorem 7.2 that $(1 - \alpha_k) v^{\alpha_k}(i, x) \to \lambda^*$ for a subsequence $\{\alpha_k\}_{k=1}^{\infty}$ converging to one. Thus, there exists an $\alpha < 1$ such that

$$(1 - \alpha) J^\alpha(i, x; U) < (1 - \alpha) v^\alpha(i, x),$$

which contradicts the definition of the value function $v^\alpha(i, x)$. □

7.6. Verification Theorem

DEFINITION 7.1 Let (λ, w) be a solution of the average optimality equation (7.18). An admissible strategy $U = (u_0, u_1, \ldots)$ is called stable with respect to w if for each initial inventory level $x \geq 0$ and for each initial demand state $i \in I$,

$$\lim_{k \to \infty} \frac{1}{k} \mathsf{E}(w(i_k, x_k)) = 0,$$

where x_k is the inventory level in period k corresponding to the initial state (i, x) and the strategy U.

LEMMA 7.7 *Let Assumptions* (i)-(ix) *hold. Then, there are constants* $S_i < \infty$ *and* $0 \leq s_i \leq S_i$, $i \in I$, *such that*

$$u^*(i,x) = \begin{cases} S_i - x, & x < s_i, \\ 0, & x \geq s_i, \end{cases}$$

attains the minimum on the RHS in (7.18) *for* $w = w^*$, *as defined in Theorem* 7.2. *Furthermore, the stationary feedback strategy* $U^* = (u^*, u^*, \ldots)$ *is stable with respect to any continuous function* w.

Proof. Let $\{\alpha_k\}_{k=0}^{\infty}$ be the sequence defined in Theorem 7.2. Let

$$G^{\alpha_k}(i,y) = c_i y + \mathsf{E}q(i, y - \xi^i) + \alpha_k F(w^{\alpha_k})(i,y) \tag{7.22}$$

and

$$G(i,y) = c_i y + \mathsf{E}q(i, y - \xi^i) + F(w^*)(i,y). \tag{7.23}$$

Because $w^*(i, \cdot)$ is K-convex, we know that a minimizer in (7.18) is given by

$$u^*(i,x) = \begin{cases} S_i - x, & x < s_i, \\ 0, & x \geq s_i, \end{cases}$$

where $0 \leq S_i \leq \infty$ minimizes $G(i, \cdot)$, and s_i solves

$$G(s_i) = K + G(S_i)$$

if a solution to this equation exists, or $s_i = 0$ otherwise. Note that if $s_i = 0$, it follows that $u^*(i,x) = 0$ for all nonnegative x. It remains to show that $S_i < \infty$.

We distinguish two cases.

- Case 1. If there is a subsequence, still denoted by $\{\alpha_k\}_{k=0}^{\infty}$, such that $s_i^{\alpha_k} > 0$ for all $k = 0, 1, \ldots$, then it follows from Lemma 7.3 that G^{α_k} attains its minimum in $[0, C_2]$ for all $\alpha_k > \alpha_0$. Thus, G^{α_k}, $k = 0, 1, \ldots$ are locally uniformly continuous and converge uniformly to G. Therefore, G attains its minimum also in $[0, C_2]$, which implies $S_i \leq C_2$.

- Case 2. If there is no such sequence, then there is a sequence, still denoted by $\{\alpha_k\}_{k=0}^{\infty}$, such that $s_i^{\alpha_k} = 0$ for all $k = 0, 1, \ldots$. It follows that for all $y > x$,
$$G^{\alpha_k}(x) < K + G^{\alpha_k}(y),$$

and therefore, in the limit,

$$G(x) < K + G(y).$$

This implies that the infimum in (7.18) is attained for $u^*(i,x) \equiv 0$, which is equivalent to $s_i = 0$. But if $s_i = 0$, we can choose S_i arbitrarily, say $S_i = C_2$.

It is obvious that the stationary policy U^* is stable with respect to any continuous function, since we have $x_k \in [0, \max\{C_2, x_0\}]$ for all $k = 0, 1, \ldots$ for such a policy. □

THEOREM 7.3 (Verification Theorem)
(a) *Let $(\lambda, w(\cdot,\cdot))$ be a solution of the average cost optimality equation* (7.18)*, with w continuous on $[0,\infty)$. Then, $\lambda \leq J(i,x;U)$ for any admissible U.*

(b) *Suppose there exists an $\hat{u}(i,x)$ for which the infimum in* (7.18) *is attained. Furthermore, let $\hat{U} = (\hat{u}, \hat{u}, \ldots)$, the stationary feedback policy given by $\hat{u}$, be stable with respect to w. Then,*

$$\begin{aligned}\lambda &= J(i,x;\hat{U}) = \lambda^* \\ &= \lim_{N\to\infty} \frac{1}{N} \mathsf{E}\left(\sum_{k=0}^{N-1} f(i_k, \hat{x}_k) + q(i_k, x_k + u_k - \xi_k) + c(i_k, \hat{u}_k) \right),\end{aligned}$$

and $\hat{U}$ is an average optimal strategy.

(c) *Moreover, $\hat{U}$ minimizes*

$$\liminf_{N\to\infty} \frac{1}{N} \mathsf{E}\left(\sum_{k=0}^{N-1} f(i_k, x_k) + q(i_k, x_k + u_k - \xi_k) + c(i_k, u_k) \right)$$

over the class of admissible decisions which are stable with respect to w.

Proof. We start by showing that

$$\lambda \leq J(i,x;U) \text{ for any } U \text{ stable with respect to } w. \tag{7.24}$$

We assume that U is stable with respect to w, and then follow the same approach used in deriving (2.15) to obtain

$$\mathsf{E}\{w(i_{k+1}, x_{k+1}) \,|\, i_0, \ldots, i_k, \xi_0, \ldots, \xi_{k-1}\} = F(w)(i_k, x_k + u_k) \quad \text{a.s.} \tag{7.25}$$

Because u_k does not necessarily attain the infimum in (7.18), we have

$$\begin{aligned} w(i_k, x_k) + \lambda \ &\leq \ f(i_k, x_k) + c(i_k, u_k) \\ &\quad + q(i_k, x_k + u_k - \xi_k) + F(w)(i_k, x_k + u_k) \quad \text{a.s.}, \end{aligned}$$

and from (7.25) we derive

$$\begin{aligned} w(i_k, x_k) + \lambda \ &\leq \ f(i_k, x_k) + q(i_k, x_k + u_k - \xi_k) + c(i_k, u_k) \\ &\quad + \mathsf{E}(w(i_{k+1}, x_k + u_k - \xi_k) \,|\, i_k) \quad \text{a.s.} \end{aligned}$$

By taking the expectation of both sides, we obtain

$$\begin{aligned} \mathsf{E}(w(i_k, x_k)) + \lambda \ &\leq \ \mathsf{E}(f(i_k, x_k) + q(i_k, x_k + u_k - \xi_k) + c(i_k, u_k)) \\ &\quad + \mathsf{E}(w(i_{k+1}, x_{k+1})). \end{aligned}$$

Summing from 0 to $n-1$ yields

$$\begin{aligned} n\lambda \ &\leq \ \mathsf{E}\left(\sum_{k=0}^{n-1} f(i_k, x_k) + q(i_k, x_k + u_k - \xi_k) \right. \\ &\quad \left. + c(i_k, u_k) \right) + \mathsf{E}(w(i_n, x_n)) - \mathsf{E}(w(i_0, x_0)). \qquad (7.26) \end{aligned}$$

Divide by n, let n go to infinity, and use the fact that U is stable with respect to w, to obtain

$$\begin{aligned} \lambda \ &\leq \ \liminf_{n \to \infty} \frac{1}{n} \mathsf{E}\left(\sum_{k=0}^{n-1} f(i_k, x_k) \right. \\ &\quad \left. + \, q(i_k, x_k + u_k - \xi_k) + c(i_k, u_k) \right). \qquad (7.27) \end{aligned}$$

Note that if the above inequality holds for 'liminf', it certainly also holds for 'limsup'. This proves (7.24).

On the other hand, if there exists a $\hat{u}$ for which the infimum in (7.18) is attained , we then have

$$\begin{aligned} w(i_k, \hat{x}_k) + \lambda \ &= \ f(i_k, \hat{x}_k) + q(i_k, \hat{x}_k + \hat{u}_k - \xi_k) + c(i_k, \hat{u}(i_k, \hat{x}_k)) \\ &\quad + F(w)(i_k, \hat{x}_k + \hat{u}(i_k, \hat{x}_k)), \quad \text{a.s.}, \end{aligned}$$

and we analogously obtain

$$\begin{aligned} n\lambda \ &= \ \mathsf{E}\left(\sum_{k=0}^{n-1} f(i_k, \hat{x}_k) + q(i_k, \hat{x}_k + \hat{u}_k - \xi_k) + c(i_k, \hat{u}(i_k, \hat{x}_k)) \right) \\ &\quad + \mathsf{E}(w(i_n, \hat{x}_n)) - \mathsf{E}(w(i_0, \hat{x}_0)). \qquad (7.28) \end{aligned}$$

Because $\hat{U}$ is assumed stable with respect to w, we get

$$\begin{aligned}\lambda &= \lim_{n\to\infty}\frac{1}{n}\mathsf{E}\left(\sum_{k=0}^{n-1} f(i_k,\hat{x}_k)+q(i_k,\hat{x}_k+\hat{u}_k-\xi_k)+c(i_k,\hat{u}(i_k,\hat{x}_k)\right)\\ &= J(i,x;\hat{U}).\end{aligned} \tag{7.29}$$

For the special solution (λ^*, w^*) defined in Theorem 7.2, and the strategy U^* defined in Lemma 7.7, we have

$$\lambda^* = J(i,x;U^*).$$

Since U^* is stable with respect to any continuous function by Lemma 7.7, it follows that

$$\lambda \le J(i,x;U^*) = \lambda^*, \tag{7.30}$$

which, in view of Lemma 7.6, proves part (a) of the theorem.

Part (a) of the theorem, together with (7.29), proves the average optimality of $\hat{U}$ over all admissible strategies. Furthermore, since $\lambda = J(i,x;\hat{U}) \ge \lambda^*$ by (7.29) and Lemma 7.6, it follows from (7.30) that $\lambda = \lambda^*$, and the proof of Part (b) is completed.

Finally, Part (c) immediately follows from Part (a) and (7.27). □

REMARK 7.5 It should be obvious that any solution (λ, w) of the average cost optimality equation and control u^* satisfying (a) and (b) of Theorem 7.3 will have a unique λ, since it represents the minimum average cost. On the other hand, if (λ, w) is a solution, then $(\lambda, w+c)$, where c is any constant, is also a solution. For the purpose of this chapter, we do not require w to be unique up to a constant. If w is not unique up to a constant, then u^* may not be unique. We also do not need w^* in Theorem 7.2 to be unique.

The final result of this section, namely, that there exists an average optimal policy of (s, S)-type, is an immediate consequence of Lemma 7.7 and Theorem 7.3.

THEOREM 7.4 *Let Assumptions* (i)-(viii) *hold. Let s_i and S_i, $i \in I$ be defined as in Lemma 7.7. Then, the stationary feedback strategy $U^* = (u^*, u^*, \dots)$ defined by*

$$u^*(i,x) = \begin{cases} S_i - x, & x < s_i,\\ 0, & x \ge s_i,\end{cases}$$

is average optimal.

7.7. Concluding Remarks and Notes

This chapter is based on Cheng and Sethi (1999b) and Beyer and Sethi (2005). We have proved a verification theorem for the average cost optimality equation, which we have used to establish the existence of an optimal state-dependent (s, S) policy.

As with the discounted cost models, the optimality of an (s, S) policy for a lost sales case is established only under the condition of zero ordering leadtime; (see references in Chapter 4). With nonzero leadtimes, the results for models with backlog do not generalize to the lost sales case. Specifically, an (s, S) policy is no longer optimal, and the form of the optimal policy is more complicated; (see, e.g., Zipkin (2008a) and Huh *et al.* (2008)). Nevertheless, an (s, S)-type policy is often used without the optimality proof; (see, e.g., Kapalka, *et al.* (1999)).

IV

MISCELLANEOUS

Chapter 8

MODELS WITH DEMAND INFLUENCED BY PROMOTION

8.1. Introduction

This chapter deals with a stochastic inventory model in which the probability distribution of the product demand in any given period depends on some environmental factors, as well as on whether or not the product is promoted in the period. The problem is to obtain optimal inventory ordering and product promotion decisions jointly so as to maximize the total profit.

Such problems arise in many business environments, where marketing tools such as promotions are often used to stimulate consumer demand. The coordination between marketing and inventory management becomes critical in their decision making. Traditionally, marketing is mainly concerned with satisfying customers while manufacturing is primarily interested in production efficiency. Conflicts may arise between the two business functions because of their different primary focus areas. Furthermore, it is necessary to evaluate the trade-off between the benefit brought by higher sales, and the increased costs caused by promotion and holding more inventory. On the other hand, a promotion plan must be supported by a coordinated procurement plan to ensure that sufficient stock is available to meet the stimulated demand.

It is obviously desirable to adopt an integrated decision making system by considering the two types of decisions jointly. However, most works in inventory literature assume exogenous demand and do not deal with promotion decisions explicitly. Early efforts in integrating promotional decisions have been mostly focused on pricing.

A typical approach for modeling the price demand relationship is to assume that the price-dependent portion of the demand is either additive

D. Beyer et al., *Markovian Demand Inventory Models*, International Series in Operations Research & Management Science 108,
DOI 10.1007/978-0-387-71604-6_8,

or multiplicative to the base demand; (see Young (1978)). Karlin and Carr (1962), Mills (1962), and Zabel (1970) studied the price demand models with this assumption, and obtained structural results for the additive and/or multiplicative cases. Furthermore, Thowsen (1975) showed that the optimal pricing/inventory policy is a base-stock list price policy under some additional assumptions. That is, if the initial inventory level is below the base-stock level, then that stock level is replenished and the list price is charged; if the initial inventory level is above the base-stock level, then nothing is ordered, and a price discount is offered.

A dynamic inventory/advertising model with partially controlled additive demand is analyzed by Balcer (1983), who obtains a joint optimal inventory and advertising strategy under certain restrictions. However, in the Balcer model, the demand/advertising relationship is deterministic because only the deterministic component of demand is affected by the advertising decision.

Sogomonian and Tang (1993) formulate a mixed integer program for an integrated promotion and production decision problem, and devise a "longest path" algorithm for finding an optimal joint promotion and production plan. Their model assumes that demand at each period depends on the time elapsed since the last promotion, the level of the last promotion, and the retail price of the current period.

In this chapter, we analyze the joint promotion/inventory management problem for a single item in the context of Markov decision processes (MDP). Specifically, we assume that consumer demand is affected by a Markov process that depends on promotion decisions. The state variable of the Markov process represents the demand state brought about by changing environmental factors as well as promotion decisions. The demand state in a period, in turn, determines the distribution of the random demand in that period.

At the beginning of each period, a decision maker observes the demand state and the inventory level, and then decides on (a) whether or not to promote and (b) how much to order. The MDP is modeled so that demand, and hence the revenue, will increase in the following period after the product is promoted. However, there is a fixed cost for promoting the product.

We solve the finite horizon problem via a dynamic programming approach. We show that there is a threshold inventory level P for each demand state such that if the threshold is exceeded, then it is desirable to promote the product. For the linear ordering cost case, the optimal inventory replenishment policy is a base-stock type policy.

Our model differs from the existing models in the literature in several ways. First, we do not stipulate a deterministic functional form of the relationship between promotion and demand. This allows the maximum flexibility for modeling the demand uncertainty in the presence of promotions. Second, we model the demand as an MDP, which provides a better modeling approach for demand under the influence of marketing activities and other uncertain environmental factors.

We should note that Sethi and Zhang (1994,1995) have developed a general stochastic production/advertising model with unreliable machines and random demand influenced by the level of advertising via a Markov process. They developed the dynamic programming equations and show how various hierarchical solutions of the problem can be constructed. They do not study the nature of optimal solutions. Their focus is on proving that the constructed hierarchical solutions are approximately optimal.

The Markov demand modeling technique adopted in this chapter extends the existing Markov-modulated demand models by introducing promotion decisions. It provides a new, flexible approach to model demand that depends not only on uncertain environmental factors, but is also influenced by promotion decisions. While we limit the scope of our analysis by including only the promotion decision, other factors that potentially affect customer demand could also be incorporated in the model in a similar way.

The remainder of the chapter is organized as follows. In Section 8.2, we present the mathematical formulation of the promotion/inventory problem, and provide a preliminary analysis of the problem in a general setting. The assumptions required for obtaining additional structural results of the model are discussed in Section 8.3. In Section 8.4, we provide the optimality proof for the optimal promotion/inventory policies. Solutions with simplified parameters are also discussed. Further extensions of the model, including those with price discounts as a promotion device, are briefly discussed in Section 8.5. A numerical study is presented in Section 8.6. The chapter concludes with some concluding remarks and end notes in Section 8.7.

8.2. Formulation of the Model

In most classical inventory models, it is usually assumed that both the price and the demand of the product are parameters and not decision variables. Hence, the revenue from sales is independent of any decision variables and can be neglected from the model except for the situation in which demand not satisfied directly from inventory cannot be backlogged. Furthermore, even in this lost sales case, the situation can be handled by including the loss in revenue as a penalty cost of the unsatisfied demand. Therefore, an optimal inventory policy can be determined by minimizing the total cost. However, in our model, promotion decisions will affect the future demand, and hence the future revenue. Thus, it is important to include the revenue explicitly in the objective function of the model.

When both revenues and costs are included in the objective function, one usually formulates a profit maximization problem. Nevertheless, in order to preserve the similarity with the approach commonly used in classical inventory literature, we will formulate the problem as a cost minimization problem by treating revenues as negative costs. This is equivalent to formulating the problem as a profit maximization problem.

8.2.1 Notation

We will formulate a discrete-time N-period problem consisting of period $0, 1, 2, \ldots, N-1$. For the temporal conventions used in our discrete-time setting, (see Figure 2.1 of Chapter 2), we introduce the following notation.

m_k = the promotion decision in period k such that
$$m_k = \begin{cases} 1, & \text{if the product is promoted in period } k \\ 0, & \text{otherwise;} \end{cases}$$

$p_{ij}(m)$ = the transition probability that the demand state changes from i to j in one period if the promotion value is m;

$\{i_k\}$ = a discrete MDP with the transition matrix $P = \{p_{ij}(m_k)\}$;

ξ_k = the demand in period k; it depends on i_k but not on k, and is independent of past demand states and past demands;

$\varphi_i(\cdot)$ = the conditional density function of ξ_k when $i_k = i$;

$\Phi_i(\cdot)$ = the distribution function corresponding to φ_i;

μ_i = $\mathsf{E}\{\xi_k | i_k = i\}$;

u_k = the nonnegative order quantity in period k;

x_k = the inventory level at the *beginning* of period k;

c_k = the unit purchase cost in period k, $k \in \langle 0, N-1 \rangle$;

c_N = the shortage cost per unit in period N;

r_k = the unit revenue in period k, $k \in \langle 0, N-1 \rangle$;

r_N = the salvage cost per unit in period N;

A_k = the promotion cost in period k when the product is promoted in the period;

h_k = the unit inventory holding cost in period k assessed on the ending inventory in the period;

p_k = the unit backlogging cost in period k assessed on the backlog at the end of the period;

$l_k(z)$ = $h_k z^+ + p_k z^-$, the inventory/backlog cost function in period k when z is the ending inventory in the period, $z^+ = \max(0, z)$, and $z^- = -\min(0, z)$.

REMARK 8.1 All of the cost/revenue parameters are assumed to be independent of the demand states for the sake of simplicity in exposition. Given that the unit revenue is independent of the promotion decision, this may imply that the model does not allow price discount. However, as it will be shown in Section 8.5, the price discount can easily be incorporated in the model by treating it as a part of the promotion cost.

8.2.2 An MDP Formulation

We suppose that an order is placed at the beginning of a period and delivered instantaneously. Also the promotion decision is made at the beginning of the period. Subsequently, the actual demand materializes during the period. The unsatisfied portion of the demand, if any, is carried forward as backlog.

The expected one-period inventory cost for period $k \in \langle 0, N{-}1\rangle$, given the demand state i, is

$$\begin{aligned} L_k(i, y) &= \mathsf{E}[l_k(y - \xi_k)|i_k = i] \\ &= \int_0^y h_k(y - \xi)\varphi_i(\xi)d\xi + \int_y^\infty p_k(\xi - y)\varphi_i(\xi)d\xi, \quad (8.1) \end{aligned}$$

where y is the amount of stock available at the beginning of a period after the order, if any, is delivered.

We assume that the revenue is received when the demand occurs. Hence, the total expected net cost in period k, given the initial inventory level x, the order quantity $u \geq 0$, and the promotion decision m, can be expressed as

$$\begin{aligned} G_k(i, x; m, u) &= \mathsf{E}[l_k(x + u - \xi_k) + A_k \mathbb{I}_{m=1} + c_k u - r_k \xi_k | i_k = i] \\ &= L_k(i, x + u) + A_k \mathbb{I}_{m=1} + c_k u - r_k \mu_i. \end{aligned}$$

In the last period N–1, if the ending inventory is positive, it is salvaged at r_N per unit; if the ending inventory is negative, the shortage is met at c_N per unit. Thus, the terminal cost function can be defined as

$$g_N(x) = c_N x^- - r_N x^+. \tag{8.2}$$

Since the unit salvage value is normally less than the purchase price, we have

$$c_N \geq r_N \geq 0. \tag{8.3}$$

Note that c_N and r_N have different meanings than c_k and r_k, $k \in \langle 0, N{-}1 \rangle$.

Let $J_n(i, x; M, U)$, $n \in \langle 0, N-1 \rangle$, denote the total expected cost, including the terminal cost, when the system is operated under the promotion policy $M = \{m_n, m_{n+1}, \ldots, m_{N-1}\}$ and the ordering policy $U = \{u_n, u_{n+1}, \ldots, u_{N-1}\}$ from period n through period $N-1$, given the initial demand state i and the inventory level x at the beginning of period n. That is,

$$J_n(i, x; M, U) = \mathsf{E}\left[\sum_{k=n}^{N-1} G_k(i_k, x_k; m_k, u_k) + g_N(x_N)\right]. \tag{8.4}$$

The inventory balance equations are given by

$$\begin{cases} x_{k+1} = x_k + u_k - \xi_k, \ k \in \langle n, N-1 \rangle, \text{ with} \\ x_n = x, \quad \text{the beginning inventory level in period } n. \end{cases} \tag{8.5}$$

The objective is to determine $M = \{m_n, m_{n+1}, \ldots, m_{N-1}\}$ and $U = \{u_n, u_{n+1}, \ldots, u_{N-1}\}$ to minimize the expected total cumulative cost. Denote $v_n(i, x)$ to be the infimum of $J_n(i, x; M, U)$, i.e.,

$$v_n(i, x) = \inf_{M,U} J_n(i, x; M, U). \tag{8.6}$$

Then, $v_n(i, x)$ satisfies the dynamic programming equations

$$\begin{cases} v_n(i, x) &= \inf_{m; u \geq 0} \{G_n(i, x; m, u) \\ & \quad + \mathsf{E}[v_{n+1}(i_{n+1}, x + u - \xi_n) \mid i_n = i]\}, \ n \in \langle 0, N-1 \rangle, \\ v_N(i, x) &= g_N(x). \end{cases} \tag{8.7}$$

Note that in this MDP formulation, the additional decision variable m takes values in a finite set. Thus, the results on the existence of optimal policies and the verification theorem can be obtained in a fashion similar to those in Chapters 2 and 4. We refer the readers to Bertsekas and Shreve (1976) for further discussion on the related issues.

Now, we rewrite (8.7) as

$$v_n(i, x) = \min_{m; y \geq x} \{G_n(i, x; m, y - x) + \mathsf{E}[v_{n+1}(i_{n+1}, y - \xi_n) \mid i_n = i]\}$$

$$\begin{aligned} &= -c_n x + \min_{m;y\geq x}\{A_n \mathbb{I}_{m=1} + c_n y + L_n(i,y) - r_n\mu_i \\ &\quad + \int_0^\infty \sum_{j=1}^L p_{ij}(m) v_{n+1}(j, y-\xi)\varphi_i(\xi)d\xi\}, \\ &\qquad n = 0,\ldots,N-1, \end{aligned} \tag{8.8}$$

$$v_N(i,x) = g_N(x). \tag{8.9}$$

To simplify our analysis, we convert (8.8) and (8.9) to an alternative set of dynamic programming equations in terms of functions w_n, using the relation

$$v_n(i,x) = w_n(i,x) - c_n x, \quad n = 0,\ldots,N. \tag{8.10}$$

The resulting dynamic programming equations are

$$\begin{aligned} w_n(i,x) &= \min_{m;y\geq x}\{A_n \mathbb{I}_{m=1} + (c_{n+1} - r_n)\mu_i + (c_n - c_{n+1})y \\ &\quad + L_n(i,y) + \int_0^\infty \sum_{j=1}^L p_{ij}(m) w_{n+1}(j, y-\xi)\varphi_i(\xi)d\xi\}, \\ &\qquad n = 0,\ldots,N-1, \end{aligned} \tag{8.11}$$

$$w_N(i,x) = g_N(x) + c_N x. \tag{8.12}$$

Let us define for $n = 0,\ldots,N-1$,

$$\begin{aligned} g_n(i,y,m) &= (c_{n+1} - r_n)\mu_i + (c_n - c_{n+1})y + L_n(i,y) \\ &\quad + \int_0^\infty \sum_{j=1}^L p_{ij}(m) w_{n+1}(j, y-\xi)\varphi_i(\xi)d\xi, \end{aligned} \tag{8.13}$$

$$q_n(i,x,m) = A_n \mathbb{I}_{m=1} + \min_{y\geq x} g_n(i,y,m). \tag{8.14}$$

Then, we have for $n = 0,\ldots,N-1$,

$$w_n(i,x) = \min_m\{q_n(i,x,m)\}. \tag{8.15}$$

For $n = N$, we use (8.2) and (8.12) and write

$$w_N(i,x) = g_N(x) + c_N x = \begin{cases} 0, & x < 0, \\ (c_N - r_N)x, & x \geq 0. \end{cases} \tag{8.16}$$

8.2.3 The Newsvendor Problem – a Myopic Solution

The value function $w_n(i,x)$ given by (8.11) and (8.12), consists of two components – the cost incurred in the current period and the cost to go. A myopic solution can be obtained by ignoring the cost to go, i.e., by ignoring the last term in (8.11). Since a promotion affects only the demand in the future, it will always be true that $m = 0$ in a myopic solution. The remaining problem becomes a newsvendor-type problem, in which the only decision variable is the inventory position y (after the order is delivered). Let us denote the cost function of the newsvendor problem by

$$\begin{aligned}\min_y g_n^b(i,y) &= (c_{n+1} - r_n)\mu_i + (c_n - c_{n+1})y \\ &\quad + L_n(i,y), \quad n \in \langle 0, N-2 \rangle. \end{aligned} \tag{8.17}$$

We can obtain the minimum of $g_n^b(i,y),\ n \in \langle 0, N-2 \rangle$, at

$$\bar{S}_{n,i} = \Phi_i^{-1}\left(\frac{c_{n+1} - c_n + p_n}{h_n + p_n}\right), \tag{8.18}$$

provided the various unit costs satisfy the condition

$$-p_n \leq c_{n+1} - c_n \leq h_n. \tag{8.19}$$

When $c_{n+1} - c_n + p_n < 0$, we have $\bar{S}_{n,i} = 0$. In this case, the "newsvendor" is always better off by postponing the purchase since the saving of $c_n - c_{n+1}$ from the decreased purchase cost outweighs the backlogging cost p_n. On the other hand, when $c_{n+1} - c_n > h_n$, we have $\bar{S}_{n,i} = \infty$. This means that the "newsvendor" makes money on each unsold unit by salvaging it at c_{n+1} in period $n+1$, which is greater than the unit purchase plus holding cost. In our dynamic inventory problem, Condition (8.19) is nothing but the usual condition which rules out any speculative motive. Henceforth, we will assume that the costs satisfy (8.19).

It is clear from Assumption 8.2 (i) that

$$\bar{S}_{n,i'} \geq \bar{S}_{n,i}, \quad \forall n \in \langle 0, N-1 \rangle \text{ and } i' > i. \tag{8.20}$$

We further assume that

$$\bar{S}_{n+1,i} \geq \bar{S}_{n,i}, \quad n \in \langle 0, N-2 \rangle, \text{ and } i \in \mathbb{I}. \tag{8.21}$$

Assumption (8.21) is equivalent to assuming that $(c_{n+1} - c_n + p_n)/(h_n + p_n)$ is nondecreasing in n. When $c_{n+1} = c_n,\ n \in \langle 0, N-1 \rangle$, (8.21) would imply that h_n/p_n is nonincreasing in n.

8.2.4 Joint Optimal Promotion and Inventory Policies

In this section, we will characterize the optimal policies derived from the dynamic programming equations by identifying the special structures that an optimal policy exhibits under certain conditions.

We will first define the optimal policy for the inventory/promotion problem as follows.

THEOREM 8.1 *The optimal inventory policy for the problem defined by* (8.11) *can be characterized by two base-stock levels, and the optimal promotion policy is of multithreshold type.*

The proof of this theorem is given in Section 8.4, after additional assumptions have been made and the relevant preliminary results are provided in Section 8.3.

The first structural property of the optimal policy is like that of the standard inventory models with linear ordering costs; a base-stock policy is still optimal. Moreover, the optimal base-stock level in each period is contingent on the demand state and the promotion decision to be made. The two base-stock levels correspond to two different decisions on promotions, respectively.

While the optimality of a base-stock policy with the base-stock level given by $S^m_{n,i}$, immediately follows from the quasi-convexity of $g_n(i,x,m)$ (the proof will be provided in Section 8.4), the promotion policy is determined by the relative position of the two functions $q_n(i,x,0)$ and $q_n(i,x,1)$. We define the difference between the two functions as

$$\Delta q_n(i,x) = q_n(i,x,0) - q_n(i,x,1), \quad n = 0,\ldots,N-1.$$

Given i and x, if $\Delta q_n(i,x) \geq 0$, then it is optimal to promote; otherwise, it is optimal not to promote.

Let us consider the equation

$$\Delta q_n(i,x) = 0. \tag{8.22}$$

Suppose there exists a finite number of real roots of this equation. Let us label these roots as $P^k_{n,i}$, $k = 1,2,\ldots,N_r$, such that $P^1_{n,i} < P^2_{n,i} < \ldots < P^{N_r}_{n,i}$. If there exists no real root of (8.22), we let $N_r = 1$ and $P^1_{n,i} = -\infty$ when $\Delta q_n(i,x) > 0$, $\forall x$, and $P^1_{n,i} = \infty$ when $\Delta q_n(i,x) < 0$, $\forall x$.

It follows that the optimal promotion policy is a function of x and is implemented in the following fashion. When $x \leq P^1_{n,i}$, do not promote; when $P^1_{n,i} < x \leq P^2_{n,i}$, promote; when $P^2_{n,i} < x \leq P^3_{n,i}$, do not promote; and so on. We may call this type of policy *a multithreshold policy.* □

8.3. Assumptions and Preliminaries

In order to provide a simple and meaningful characterization to the optimal policy for the somewhat general MDP problem formulated in Section 8.3, we introduce some additional assumptions.

8.3.1 Quasi-convexity

To obtain the optimality of simple form policies for our model, we would like to have the function $g_n(i, y, m)$ to be quasi-convex in y according to Definition C.1.2. For this purpose, we make the following important, albeit restrictive, assumption on the demand density functions.

ASSUMPTION 8.1 *The demand density functions* $\varphi_i(\cdot)$, $i = 1, \ldots, L$, *are assumed to be of Pólya frequency of order* 2 *(*PF_2*) according to Definition* C.1.1.

Assumption 8.1 is, in fact, a popular assumption made in the inventory literature; (see Porteus (1971)). As Porteus and others have argued, such a condition is not very restrictive because the class of PF_2 densities is not lacking significant members. It contains all exponential densities and all finite convolutions of such densities. Furthermore, any mean μ and any variance in the range $[\mu^2/n, \mu^2]$ can be produced with a convolution of n exponential densities.

We will further note that such a condition on demand densities is not necessary for our results. However, we use this condition for the convenience of mathematical derivation for our results.

8.3.2 Stochastic Dominance

In the inventory literature (see the review in Section 8.1), it is usually assumed that demand consists of a deterministic component and a stochastic component, with promotion activities affecting only the deterministic component. Such an assumption provides a simplified mathematical modeling approach for describing the promotion/demand relationship. However, it does not accurately reflect the real-world situation, since the effect of promotional activities on demand cannot be predetermined with certainty. To capture the stochastic nature of the promotion/demand relationship, we will use a more realistic modeling approach based on the concept of *stochastic dominance* for our joint inventory/promotion decision problem.

Modeling the demand uncertainty induced by environmental factors and promotional activities in a stochastic fashion has not been attempted in the inventory literature. Our models represent a new approach to demand modeling. Based on the stochastic ordering relations introduced above, we are able to describe the relationship between demand and promotion activities in mathematical terms. See Section B.5 for additional information on stochastic dominance and some of the commonly used definitions of stochastic ordering.

For the purpose of the analysis carried out in this chapter, we adopt Definition B.5.1 of stochastic ordering.

ASSUMPTION 8.2

(i) *There exists a stochastic ordering relation between demands in different demand states such that*

$$\zeta_1 \underset{\text{st}}{\leq} \zeta_2 \underset{\text{st}}{\leq} \cdots \underset{\text{st}}{\leq} \zeta_L.$$

(ii) *For any* $1 \leq l \leq L$,

$$\sum_{j=l}^{L} p_{ij}(1) \geq \sum_{j=l}^{L} p_{ij}(0), \text{ and} \tag{8.23}$$

(iii)

$$\sum_{j=l}^{L} p_{ij}(m) \text{ is nondecreasing in } i. \tag{8.24}$$

REMARK 8.2 In simpler terms, Assumption 8.2 (i) means that the demand in a higher demand state is more likely to be larger than that in a lower demand state. Assumption 8.2 (ii) reflects the fact that a promotion would more likely lead to a demand state with a stochastically larger demand. Assumption 8.2 (iii) means that if the current demand state is higher, the next period is more likely to be in a demand state with a stochastically larger demand. Together, these assumptions will ensure that a promotion not only always generates a stochastically greater demand in the next period, but also has a positive impact on future demand.

Assumption 8.2 may not apply to the cases in which customers build inventories during an on-sale period and purchase less afterwards. However, our current model is more suitable to the situation where promotions are made in the form of advertising.

ASSUMPTION 8.3 $p_{ij}(1) = 0, \forall j < i$ *and* $p_{ij}(0) = 0, \forall i < j$.

REMARK 8.3 Assumption 8.3 means that the demand state in the next period after the product is promoted cannot be stochastically smaller than that in the current period. Likewise, if the product is not promoted in the current period, the demand state in the next period cannot be stochastically greater than that in the current period. Assumption 8.3 is stronger than Assumption 8.2, in the sense that some of the transition probabilities are fixed at zero. However, the nonzero elements of the transition probability matrix are not binding by Assumption 8.2.

The purpose of making these assumptions is to overcome some mathematical difficulties that are involved in proving the optimality of the simple form policies introduced in Section 8.4.2.

8.4. Structural Results

We have formulated a profit-maximizing inventory/promotion decision problem as a cost minimization problem, and have developed the dynamic programming equations for it.

In Section 8.2, we have characterized the general structure of the optimal policies without detailed proofs. We have shown that the optimal policies for (8.7) can be expressed in simple forms. Specifically, the inventory policy still retains the simplicity of base-stock policies, while the promotion policy is of threshold type.

In this section, we will provide the proof for the optimality of the policies defined in Theorem 8.1. Furthermore, we will show that the optimal policies can be further simplified under certain conditions.

8.4.1 Quasi-convexity of g_n

Now we present some useful properties of the dynamic programming equations and functions involved in (8.11-8.15).

LEMMA 8.1 *Denote the minimizer of* $g_n(i, y, m)$ *by* $S^m_{n,i}$. *For* $i = 1, \ldots, L$ *and* $m = 0, 1$,

(a) $g_n(i, y, m)$ *is quasi-convex in* y *with* $S^m_{n,i} \geq 0$ *for* $n = 0, \ldots, N-1$*;*

(b) $w_n(i, x)$ *is nondecreasing in* x *and constant when* $x \leq 0$ *for* $n = 0, \ldots, N$.

Proof. In view of $c_N \geq r_N$, part(b) is obvious from (8.16) when $n = N$. It is easy to see from (8.13) and (8.15) that (b) follows directly from (a) for $n < N-1$.

We will now prove part (a) for $n = 0, \ldots, N-1$ by induction. Let us define

$$\bar{g}_n(i, y, m) = (c_n - c_{n+1})y + l_n(y) + \sum_{j=1}^{L} p_{ij}(m) w_{n+1}(j, y),$$

where $l_n(y)$ is the surplus cost function as defined in Section 8.2.1. Then, (8.13) can be written as

$$g_n(i, y, m) = (c_n - r_n)\mu_i + \int_0^\infty \bar{g}_n(i, y - \xi, m)\varphi_i(\xi)d\xi.$$

When $n = N-1$, $w_{n+1}(j, y) = w_N(j, y)$ is nondecreasing in y and is a constant when $y \leq 0$ by definition. From the definition of $l_n(y)$, we have

$$(c_n - c_{n+1})y + l_n(y) = (c_n - c_{n+1} + h_n)y^+ + (p_n - c_{n+1} + c_n)y^-,$$

which is convex in y with the minimum at 0. Therefore, $\bar{g}_n(i, y, m)$ is quasi-convex in y. Furthermore, the demand density φ_i is assumed to be PF$_2$. By Theorem C.1.1, $\int_0^\infty \bar{g}_n(i, y - \xi, m)\varphi_i(\xi)d\xi$, and hence, $g_n(i, y, m)$ is also quasi-convex in y with its minimizer $S_{n,i}^m \geq 0$. Now, we have completed the induction for the base case $n = N-1$.

Let us assume that (a) holds for $n = k < N-1$ and prove that (a) is true for $n = k-1$. Since (a) is assumed to be true for $k = n$, it follows that (b) is true for $k = n$, i.e., $w_{n+1}(j, y)$ is nondecreasing in y and constant when $y \leq 0$. Using the same line of reasoning used above for the base case, we can show that for $n = k-1$, $g_n(i, y, m)$ is quasi-convex in y with its minimizer $S_{n,i}^m \geq 0$. Therefore, (a) is proved for $n = k-1$, and the induction is completed. □

8.4.2 Simple Form Solutions

In this subsection we will demonstrate that under certain conditions, the optimal solution exhibits some special structures that will simplify its computation greatly. Furthermore, the simplicity of the special structures allows for easier interpretations and implementation of optimal policies. The simple form solutions will be obtained with additional assumptions made about demand and transition probabilities of the Markov chain.

8.4.2.1 (S^0, S^1, P) Policies

We will show in this subsection that if a stochastic dominance relationship exists between the demands with and without promotions, the optimal promotion and ordering policy for the finite horizon problem can be characterized as an (S^0, S^1, P) *policy*, which is defined below.

DEFINITION 8.1 An (S^0, S^1, P) policy is specified by three control parameters S^0, S^1, and P with $S^0 \leq S^1$. Under an (S^0, S^1, P) policy, the product is promoted when the initial inventory x is above or equal to P. Furthermore, an order is placed to increase the inventory position to S^1

when $x < S^1$ and the product is promoted, and to S^0 when $x < S^0$ and the product is not promoted.

This is a special case of the optimal policies described in Section 8.4. Here, the multithreshold policy for promotion decisions is simplified to a single threshold policy.

First, we show some monotonicity properties that are needed for deriving the new results.

LEMMA 8.2 *Let $c_N = r_N$ and let Assumption 8.2 hold. Then for $1 \le i < i' \le L$,*
(a) $w_n(i, x) \ge w_n(i', x)$, $n \in \langle 0, N\rangle$;
(b) $S^1_{n,i'} \ge S^1_{n,i}$, $n \in \langle 0, N-1\rangle$; *and*
(c) $S^1_{n+1,i} \ge S^1_{n,i}$ *and* $S^1_{n,i} \ge S^0_{n,i}$, $n \in \langle 0, N-1\rangle$.

Proof. We prove (a) by induction. When $n = N$, $w_N(i, x) = 0$ by definition. Suppose that $w_{n+1}(i, x)$ is nonincreasing in i. First, let us prove that $g_n(i, y, m) \ge g_n(i', y, m), \forall m, y$. Since $w_{n+1}(j, y)$ is nonincreasing in j, according to Assumption 8.2 (iii), $\sum_{j=1}^{L} p_{ij}(m)w_{n+1}(j, y)$ is also nonincreasing in i. Furthermore, by (B.3) we have

$$\int_0^\infty \sum_{j=1}^{L} p_{ij}(m)w_{n+1}(j, y-\xi)\varphi_i(\xi)d\xi \ge \int_0^\infty \sum_{j=1}^{L} p_{ij}(m)w_{n+1}(j, y-\xi)\varphi_{i'}(\xi)d\xi.$$

Using Assumption 8.2 (ii), we obtain that

$$\int_0^\infty \sum_{j=1}^{L} p_{ij}(m)w_{n+1}(j, y-\xi)\varphi_{i'}(\xi)d\xi \ge \int_0^\infty \sum_{j=1}^{L} p_{i'j}(m)w_{n+1}(j, y-\xi)\varphi_{i'}(\xi)d\xi.$$

On account of

$$g_n(i, y, m) = g^b_n(i, y) + \int_0^\infty \sum_{j=1}^{L} p_{ij}(m)w_{n+1}(j, y-\xi)\varphi_i(\xi)d\xi \quad (8.25)$$

and (8.13), it is clear that

$$g_n(i, y, m) \ge g_n(i', y, m).$$

In view of (8.14) and the quasi-convexity of g_n, we have

$$q_n(i,x,m) \geq q_n(i',x,m), \forall i' > i, x, m.$$

Finally, from (8.15), we conclude that

$$w_n(i,x) \geq w_n(i',x),$$

which proves (a).

Let us examine (8.25) again. For $n = N-1$, it is clear that $S^1_{N-1,i} = \bar{S}_{N-1,i}$ since $w_N(i,x) = 0$. Hence, from (8.20) we know that $S^1_{N-1,i'} \geq S^1_{N-1,i}, \forall i' > i$. When $n = N-2$, from Assumption 8.3 we have

$$\begin{aligned} & \int_0^\infty \sum_{j=1}^{L} p_{ij}(1) w'_{n+1}(j, y-\xi)\varphi_i(\xi)d\xi \\ = & \int_0^\infty \sum_{j=i}^{L} p_{ij}(1) w'_{n+1}(j, y-\xi)\varphi_i(\xi)d\xi, \end{aligned}$$

which is equal to 0 when $y \leq \bar{S}_{N-1,i}$ and is nonnegative otherwise, according to Lemma 8.1 (b). By Assumption (8.21), $\bar{S}_{N-2,i} \leq \bar{S}_{N-1,i}$. Thus, $S^1_{N-2,i} = \bar{S}_{N-2,i}$. Therefore, by (8.20), $S^1_{N-2,i'} \geq S^1_{N-2,i}, \forall i' > i$. Repeating this step for $n = N-3, \ldots, 0$, we complete the proof for (b).

Since $S^1_{n,i} = \bar{S}_{n,i}$, then according to (8.21), $S^1_{n+1,i} \geq S^1_{n,i}$. Now let us examine (8.25) with $m = 0$. Since

$$\int_0^\infty \sum_{j=1}^{L} p_{ij}(0) w'_{n+1}(j, y-\xi)\varphi_i(\xi)d\xi \geq 0,$$

we have $S^0_{n,i} \leq \bar{S}_{n,i}$, i.e., $S^0_{n,i} \leq S^1_{n,i}$. Thus, (c) is proved. □

REMARK 8.4 The assumption $c_N = r_N$ is used in this proof for convenience in the exposition. However, it can be relaxed by recalculating $\bar{S}_{N-1,i}$ with the last item included in (8.13).

Because of Assumption 8.3, we are able to determine the optimal base-stock level in a myopic fashion, which is reflected in part (b) of Lemma 8.2. We are now ready to show that the optimal policies for the N-period problem are of (S^0, S^1, P) type.

THEOREM 8.2 *When $x \leq S^1_{n,i}$, an $(S^0_{n,i}, S^1_{n,i}, P_{n,i})$ policy is optimal for the problem defined by* (8.10-8.13) *under Assumption 8.2, with $P_{n,i}$ given by the smallest root of equation* (8.22).

Proof. From the quasi-convexity of $g_n(i,y,m)$, $g'_n(i,y,m) \geq 0$ when $y \leq S^m_{n,i}$. By (8.14), we have $q'_n(i,x,m) = 0$ when $x \leq S^m_{n,i}$, and $q'_n(i,x,m) \geq 0$ when $x > S^m_{n,i}$. Also, we know that $S^0_{n,i} \leq S^1_{n,i}$ by Lemma 8.2 (c). Hence, for $x \leq S^1_{n,i}$,

$$\frac{\partial}{\partial x}(q_n(i,x,0) - q_n(i,x,1)) \geq 0,$$

which means that there exists at most one real root for (8.22) when $x \leq S^1_{n,i}$. Hence, a simple threshold policy is optimal for the promotion decision when $x \leq S^1_{n,i}$. □

REMARK 8.5 First, $x \leq S^1_{n,i}$ is not a necessary condition for Theorem 8.2. It is used for obtaining a simple proof. Second, the condition is not very restrictive and is satisfied for all $n \geq k$ if $x_k \leq S^1_{k,i}$ for any period $k \geq 0$ with demand state i. This can be seen from the following argument. By Assumption 8.3, if the product is promoted in period k, the demand state j in period $k+1$ will not be lower than i. Since $S^1_{k,i} \leq S^1_{k+1,j}$, and $\forall j \geq i$ and $S^0_{k,i} \leq S^1_{k,i}$, $\forall i$ as shown in Lemma 8.2 (c), then as soon as $x_k \leq S^1_{k,i}$ for any $k \geq 0$, we have $x_n \leq S^1_{n,i}$ for all subsequent periods $n = k+1, k+2, \ldots, N-1$ and all possible i. In view of these arguments, the condition will hold when $x_0 = 0$, which is often the case.

REMARK 8.6 Without further specifying cost and density functions, we may have three typical cases in terms of the relative position of $P_{n,i}$ as given below.

- Case 1. $P_{n,i} = -\infty$. In this case, a promotion is always desired. The corresponding optimal inventory policy is a base-stock policy with the base-stock level given by $S^1_{n,i}$. In fact, this is the only case in which $P_{n,i} < S^0_{n,i}$, since $q'_n(i,x,m) = 0$ when $x \leq S^m_{n,i}$, $m = 1, 2$.

- Case 2. $S^0_{n,i} < P_{n,i} \leq S^1_{n,i}$. In this case, a promotion is desirable only when the initial inventory level exceeds the critical level $P_{n,i}$. Two base-stock levels $S^0_{n,i}$ and $S^1_{n,i}$ are applied, respectively, depending on whether or not the product is promoted.

- Case 3. $S^0_{n,i} \leq S^1_{n,i} < P_{n,i}$. This case is similar to Case 2 except that no order will be placed once the product is promoted.

8.4.2.2 (S, P) Policies

In this subsection, we replace Assumptions 8.2 (i) and 8.2 (iii) by the following.

ASSUMPTION 8.4 *We assume that* $\xi_1 \leq \xi_2 \leq \ldots \xi_L$, *in the sense of the first moment ordering.*

REMARK 8.7 Although Assumption 8.4 is more restrictive than Assumption 8.2 (i) used in the previous subsection, it still represents a generalization of the commonly used assumption of the additive random demands in the literature. On the positive side, we can now drop Assumption 8.3 for the results to be obtained in this subsection. We further note that Assumption 8.4 can be replaced by some weaker assumptions under which the results in this subsection would still hold.

THEOREM 8.3 *Under Assumptions* 8.4, 8.2 (ii), *and* 8.2 (iii), *a special case of the* (S^0, S^1, P) *policy, i.e., the* (S, P) *policy with* $S_{n,i} = S^0_{n,i} = S^1_{n,i}$, *is optimal for the model.*

Proof. Assumption 8.4 is equivalent to

$$\varphi_1(x - \mu_1) = \varphi_2(x - \mu_2) = \cdots = \varphi_L(x - \mu_L)$$

in view of Definition B.5.3. Since ξ_1 is nonnegative, i.e., $\varphi_1(x) = 0, \forall x \leq 0$, it follows that

$$\varphi_i(x) = 0, \quad \forall x \leq \mu_i - \mu_1, \; i = 1, \ldots, L. \tag{8.26}$$

Also from (8.18),

$$\bar{S}_{n,1} - \mu_1 = \bar{S}_{n,2} - \mu_2 = \cdots = \bar{S}_{n,L} - \mu_L. \tag{8.27}$$

The proof is done by induction. For $k = N-1$, $S_{N-1,i} = S^0_{N-1,i} = S^1_{N-1,i} = \bar{S}_{N-1,i}$ for all i. Assume for $k = n+1$ that $S_{n+1,i} = S^0_{n+1,i} = S^1_{n+1,i} = \bar{S}_{n+1,i}$ for all i.

It is known from Lemma 8.1 (b) that $w'_{n+1}(i, y) = 0, \; \forall y \leq S^0_{n+1,i}$, and $w'_{n+1}(i, y) \geq 0, \; \forall y > S^0_{n+1,i}$. Thus,

$$\sum_{j=1}^{L} p_{ij}(m) w'_{n+1}(j, y) \begin{cases} = 0, & y \leq S^0_{n+1,1} = \bar{S}_{n+1,1}, \\ \geq 0, & \text{otherwise.} \end{cases}$$

Furthermore, by (8.26) and (8.27),

$$\int_0^\infty \sum_{j=1}^{L} p_{ij}(m) w'_{n+1}(j, y - \xi) \varphi_i(\xi) d\xi \begin{cases} = 0, & y \leq \bar{S}_{n+1,1} + \mu_i - \mu_1 \\ & \qquad = \bar{S}_{n+1,i}, \\ \geq 0, & \text{otherwise.} \end{cases}$$

Since $\bar{S}_{n,i} \leq \bar{S}_{n+1,i}$, we conclude that $S_{n,i} = S^0_{n,i} = S^1_{n,i} = \bar{S}_{n,i}$. □

8.5. Extensions

In this section, we briefly discuss some extensions of our model.

Infinite Horizon Problems. When the planning horizon is infinite, the formulation of the model can be presented as follows. The total cost of the infinite horizon problem to be minimized is given by

$$J_n(i,x;M,U) = \mathsf{E}\sum_{k=n}^{\infty} \alpha^{k-n} G_k(i_k, x_k; m_k, u_k), \tag{8.28}$$

where $0 < \alpha < 1$ is a discount factor. The inventory balance equations are given by

$$\begin{cases} x_{k+1} = x_k + u_k - \xi_k, \ k = n, n+1, \ldots, \text{ with} \\ x_n = x, \ \text{ the initial inventory level.} \end{cases} \tag{8.29}$$

Denote the minimum of $J_n(i,x;M,U)$ by $v_n(i,x)$, i.e.,

$$v_n(i,x) = \inf_{M,U} J_n(i,x;M,U). \tag{8.30}$$

Then, $v_n(i,x)$ satisfies the dynamic programming equations

$$\begin{aligned} v_n(i,x) &= \min_{m;u\geq 0} \{G_n(i,x;m,u) \\ &\quad +\alpha\, \mathsf{E}[v_{n+1}(i_{n+1}, x+u-\xi_n)|i_n = i]\}, \\ &\quad n = 0,1,\ldots. \end{aligned} \tag{8.31}$$

By assuming stationary data, we can suppress the time index in the formulation and expect that there exists a stationary optimal policy which does not depend on time. (Issues regarding the existence of an optimal feedback-type policy in infinite horizon models have been addressed in Bertsekas and Shreve (1976).) Therefore, the DP equation can be written as

$$\begin{aligned} v(i,x) &= \min_{m;u\geq 0} \{G(i,x;m,u) \\ &\quad +\alpha \sum_{j=1}^{L} p_{ij}(m) \int_0^\infty v(j, x+u-\xi)\varphi_i(\xi)d\xi\}. \end{aligned} \tag{8.32}$$

Since the infinite horizon problem can be considered as a limiting case of the finite horizon problem when $N \to \infty$, its optimal policy will have the same structure as the optimal policy in the finite horizon case, i.e., base-stock inventory policies and the threshold promotion policy, or (S^0, S^1, P) policy, under the corresponding assumptions made in the finite horizon problem.

Lost Sales. The dynamic programming equations can be written as

$$\begin{cases} v_n(i,x) &= \min_{m;u\geq 0} \{G_n(i,x;m,u) \\ & \quad + \int_0^\infty [v_{n+1}(i_{n+1},(x+u-\xi)^+)\varphi_i(\xi)d\xi\}, \\ & \qquad 0 \leq n \leq N-1, \\ v_N(i,x) &= -c_N x^+. \end{cases} \tag{8.33}$$

For this problem, we only need to verify that

$$g_n(i,y,m) = c_n y + L_n(i,y) + \int_0^\infty \sum_{j=1}^L p_{ij}(m) v_{n+1}(j,(y-\xi)^+)\varphi_i(\xi)d\xi$$

is also quasi-convex in y, where $L_n(i,y)$ is defined as

$$L_n(i,y) = \int_0^\infty [l_n(i,y-\xi) - r_n(y - \mathsf{E}(y-\xi)^+)]\varphi_i(\xi)d\xi.$$

Furthermore, we have

$$\begin{aligned} g_n(i,y,m) &= (c_n - r_n)y + \int_0^\infty [l_n(i,y-\xi) + r_n \mathsf{E}(y-\xi)^+ \\ & \quad + \sum_{j=1}^L p_{ij}(m) w_{n+1}(j,(y-\xi)^+) \\ & \quad - c_{n+1}\mathsf{E}(y-\xi)^+]\varphi_i(\xi)d\xi \\ &= \int_0^\infty [l_n(i,y-\xi) + (r_n - c_n)(y-\xi)^- \\ & \quad + (c_n - c_{n+1})(y-\xi)^+ \\ & \quad + \sum_{j=1}^L p_{ij}(m) w_{n+1}(j,(y-\xi)^+)]\varphi_i(\xi)d\xi. \end{aligned}$$

It is easy to verify that

$$l_n(i,y) + (r_n - c_n)y^- + (c_n - c_{n+1})y^+ + \sum_{j=1}^L p_{ij}(m) w_{n+1}(j,y^+)$$

is quasi-convex in y. Therefore, $g_n(i,y,m)$ is also quasi-convex in y since the density of ξ_i is assumed to be PF_2.

Price Discounts. Price discounts can be considered a part of the promotion cost. Let b_k be a proportional price discount offered if the product is promoted in period k. Then, the actual unit revenue will be $(1-b_k)r_k$. By considering the revenue loss as part of the promotion cost, the problem has the same form as the original problem.

Multiple Promotion Levels. We assume that the promotion effort can take on one of M discrete levels. These discrete levels could represent promotions communicated through different advertising media. At different promotion levels, the promotion expenditures incurred are also different.

Carryover Effect of Promotions. First of all, the carryover effect of a promotion can be partially captured by the Markovian transition law associated with the demand process. However, we may also incorporate this factor explicitly in the following way. If we assume that the effect of a promotion lasts $J-1$ periods, then we can redefine the demand state as

$$\tilde{i} = (i,j), \; i = 1,\dots,L, \; j = 0,\dots,J-1,$$

where j is the number of periods elapsed from the last promotion. In such a way, we can transform the new problem to the standard form and obtain a similar solution.

Nonlinear Inventory Cost Function. By assuming the demand density function to be PF_2, we may relax the requirements on the inventory cost function. We only require $l_k(x)$ to be quasi-convex in x and attain its minimum at $x = 0$. It is clear that such a generalized form of $l_k(x)$ will not change the results obtained in this chapter.

Case	c	α	p	q	A	S_1^0, S_2^0, S_3^0	S_1^1, S_2^1, S_3^1
3.1	0.50	0.90	1.00	5.00	2.00	12,12,16	16,17,17
3.2	1.00	0.90	1.00	2.00	2.00	8,8,12	12,15,15
3.3	2.00	0.90	1.00	2.00	2.00	0,0,0	0,0,0
3.4	2.50	0.90	1.00	2.00	2.00	0,0,0	0,0,0
3.5	1.00	0.90	2.00	4.00	2.00	9,9,13	13,16,16
3.6	1.00	0.90	1.00	4.00	2.00	11,11,15	15,16,16
3.7	1.00	0.90	1.00	5.00	2.00	11,11,15	15,17,17
3.8	0.50	0.90	1.00	2.00	2.00	9,9,13	13,16,16
3.9	0.50	0.90	1.00	1.00	2.00	7,7,10	10,14,14
3.10	0.50	0.95	1.00	2.00	2.00	9,9,13	13,16,16
3.11	0.50	0.85	1.00	2.00	2.00	9,9,13	13,16,16

Table 8.1. Numerical results for the MDP model with uniform demand distribution.

Case	c	α	p	q	A	S_1^0, S_2^0, S_3^0	S_1^1, S_2^1, S_3^1
4.1	0.50	0.90	1.00	5.00	2.00	8,8,14	14,19,19
4.2	1.00	0.90	1.00	2.00	2.00	4,4,10	10,15,15
4.3	2.00	0.90	1.00	2.00	2.00	0,0,0	0,0,0
4.4	2.50	0.90	1.00	2.00	2.00	0,0,0	0,0,0
4.5	1.00	0.90	2.00	4.00	2.00	6,6,12	12,17,17
4.6	1.00	0.90	1.00	4.00	2.00	7,7,13	13,18,18
4.7	1.00	0.90	1.00	5.00	2.00	7,7,13	13,18,18
4.8	0.50	0.90	1.00	2.00	2.00	6,6,12	12,17,17
4.9	0.50	0.90	1.00	1.00	2.00	3,3,9	9,14,14
4.10	0.50	0.95	1.00	2.00	2.00	6,6,12	12,17,17
4.11	0.50	0.85	1.00	2.00	2.00	6,6,12	12,17,17

Table 8.2. Numerical results for the MDP model with truncated normal demand distribution.

8.6. Numerical Results

In this section, we design a comparison study to demonstrate the advantage of the joint promotion and inventory decision making. For the same sample data, two types of policies are computed. One is the (S^0, S^1, P) policy, which we have shown to be optimal under certain conditions. The other is the policy that one would use in a decentralized system, i.e., promotion decisions and inventory decisions are made separately. More specifically, we assume that in a decentralized system, the information about the current inventory level is not considered for the promotion decision making purpose. A promotion will be conducted only if the expected revenue increase exceeds the promotion cost.

Our numerical results confirm that the joint decision making leads to better performance than the decentralized decision making. Tables 8.1 and 8.2 are the results for the cases with uniform demand distributions and truncated normal distributions, respectively. The parameters c, α, h, p and A are the unit purchase cost, the discount factor, the unit holding cost, the unit shortage cost, and the promotion cost, respectively. Figures 8.1-8.4 plot some of the cost functions obtained by two types of decision making systems corresponding to the Cases 3.1-3.4 in Table 8.1 . The solid lines (OP) are the results using our MDP models, while the dotted lines (DC) represent the results obtained based on the decentralized system.

The base-stock values computed using the two different approaches happen to be the same in these particular cases. However, the values of the cost functions are different. In each case, our model performs no worse than the decentralized model.

8.7. Concluding Remarks and Notes

In this chapter based on Cheng and Sethi (1999a) and Cheng (1996), we have developed an MDP model for a joint inventory/promotion decision problem, where the state variable of the MDP represents the demand state brought about by changing environmental factors as well as promotion decisions. Optimal inventory and promotion decision policies in a finite horizon setting are obtained via dynamic programming. Under certain conditions, we show that there is a threshold inventory level P for each demand state such that if the threshold is exceeded, then it is desirable to promote the product. For the proportional ordering cost case, the optimal inventory replenishment policy is a base-stock type policy with the optimal base-stock level dependent on the promotion decision. Several extensions of the model are also discussed. We also provided numerical results in Section 8.6 to demonstrate the benefit achieved by the joint inventory/promotion decision making.

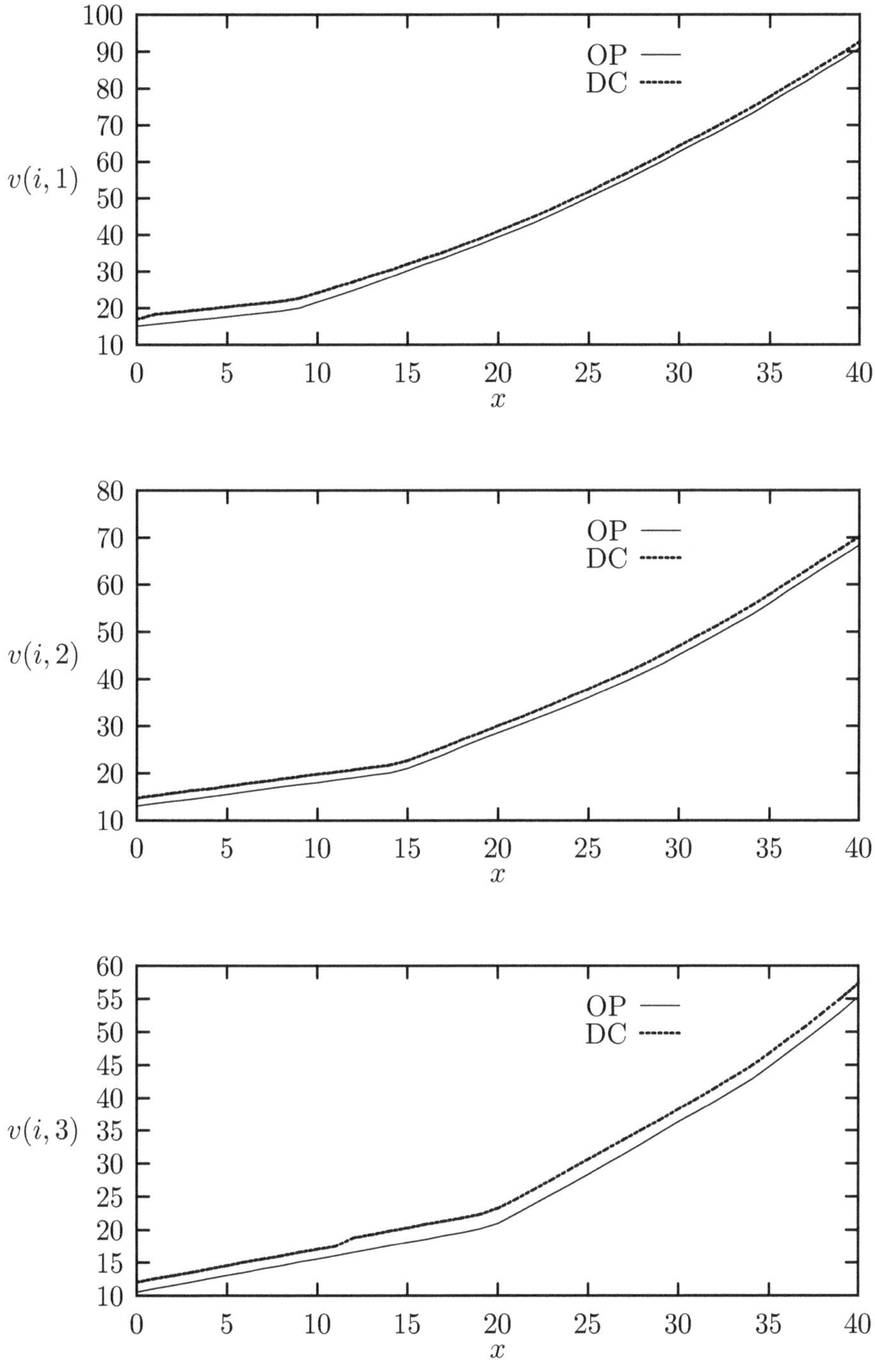

Figure 8.1. Numerical results for Case 3.1.

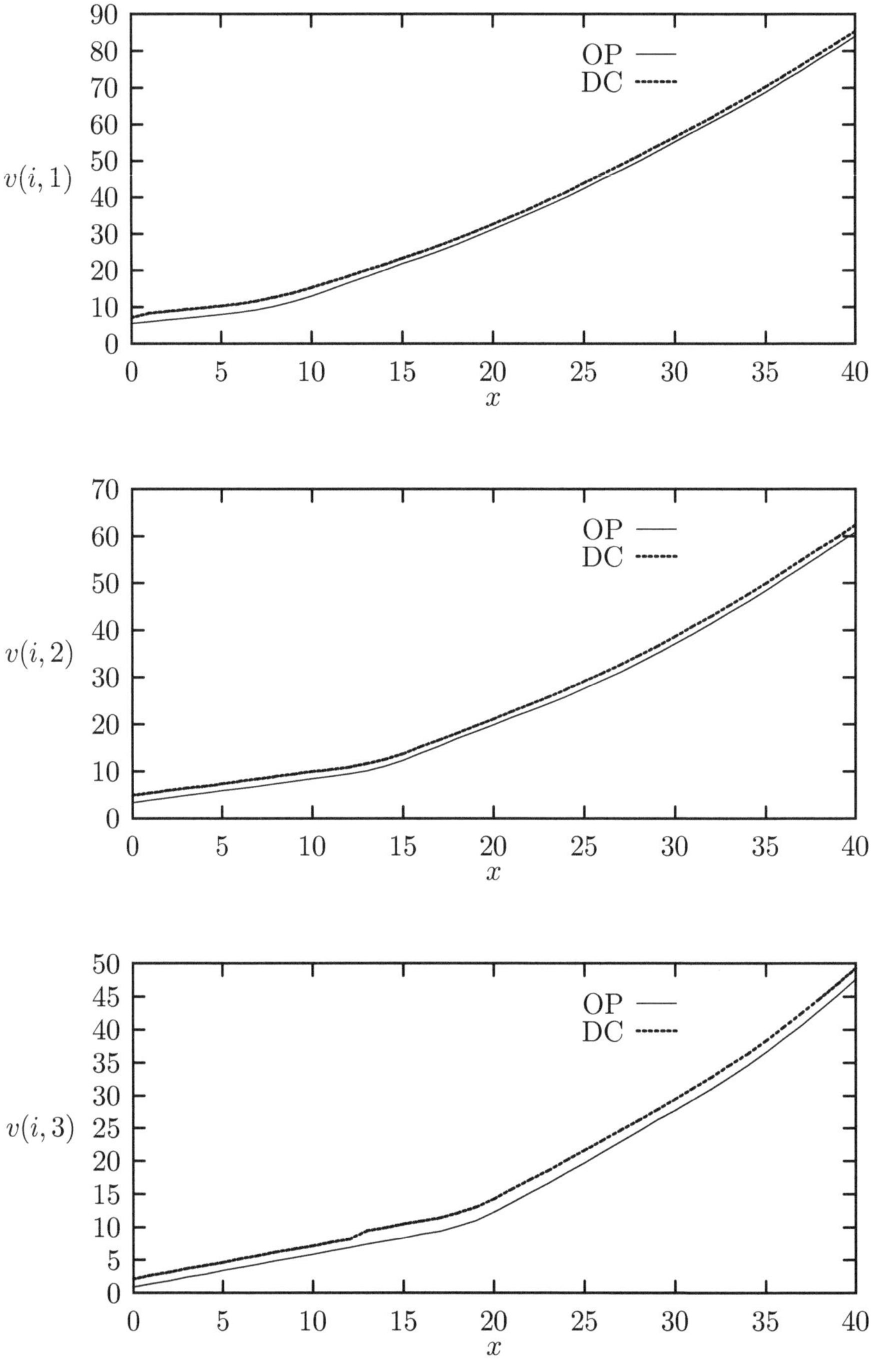

Figure 8.2. Numerical results for Case 3.2.

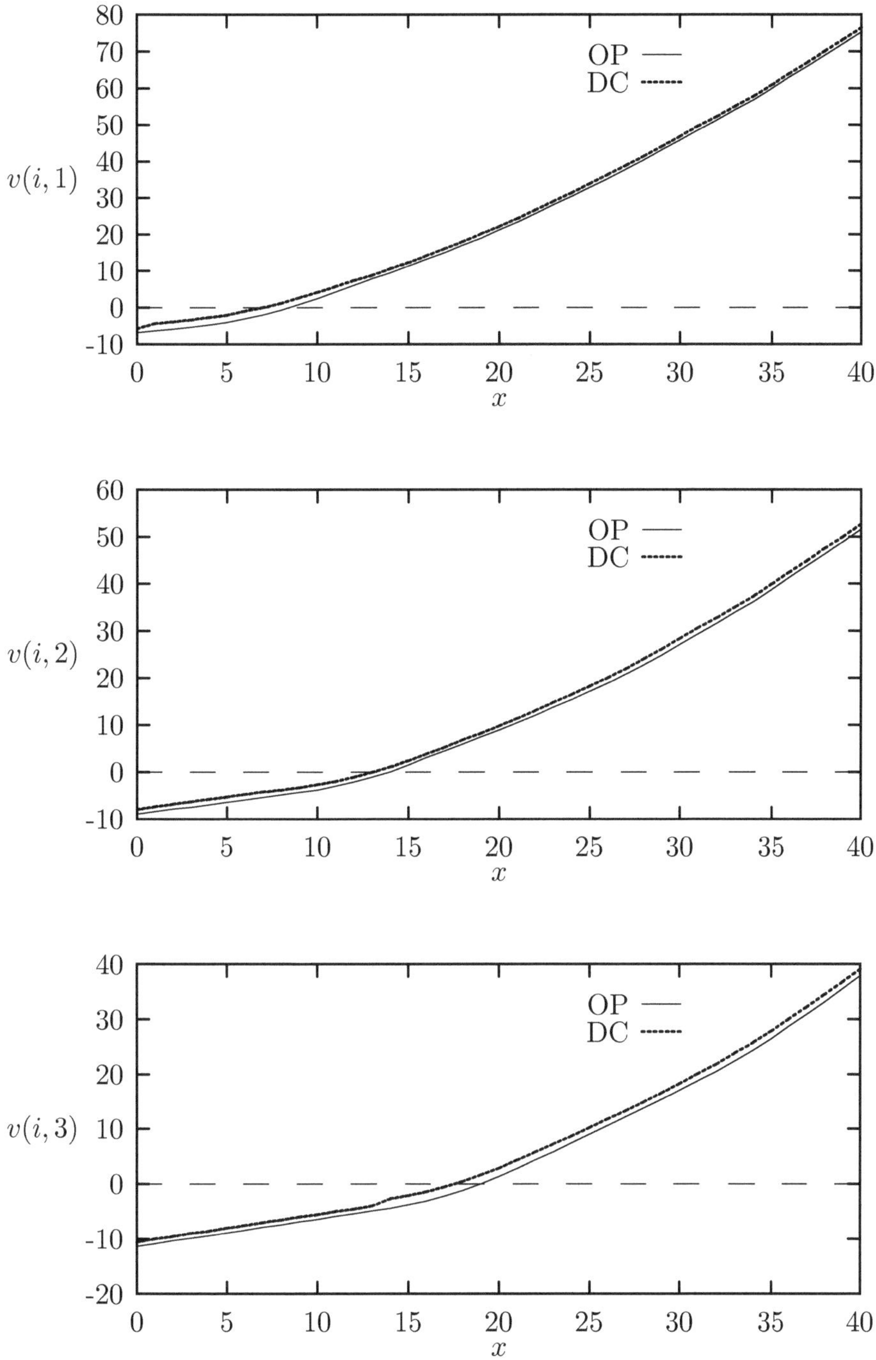

Figure 8.3. Numerical results for Case 3.3.

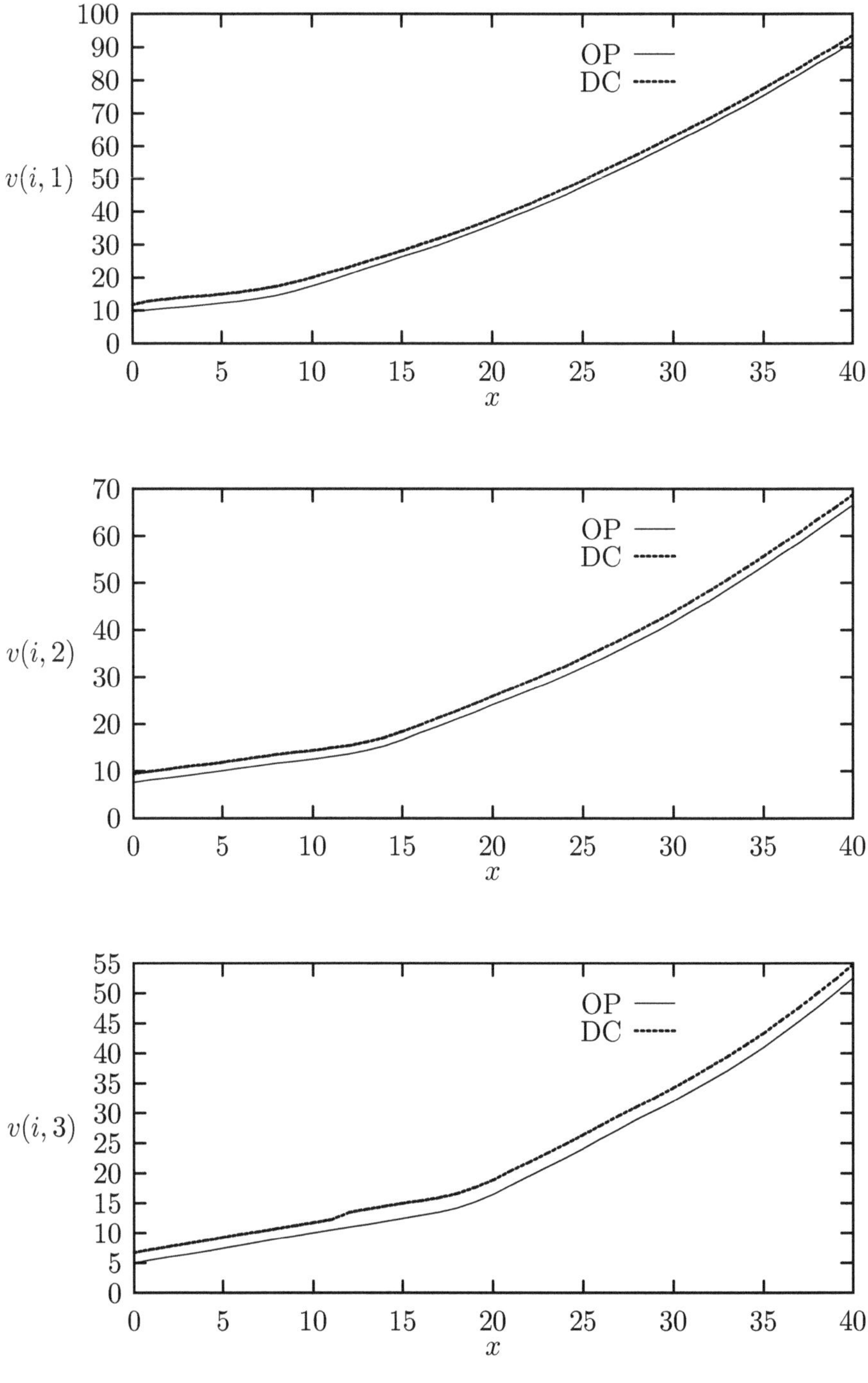

Figure 8.4. Numerical results for Case 3.4.

Chapter 9

VANISHING DISCOUNT APPROACH VERSUS STATIONARY DISTRIBUTION APPROACH

9.1. Introduction

In Part III, we derived the structure of the optimal policy to minimize the long-run average cost by using the vanishing discount method. In the classical inventory literature, a stationary distribution approach is often used to minimize the long-run average cost. In this approach, the stationary distribution of the inventory levels is obtained for a specific class of policies, the best policy in this class is found, and then it is proven that this policy is average optimal. In this chapter, we review the stationary distribution approach in solving the simpler problem of an inventory model with i.i.d demands, and then show how the results of this analysis relate to those obtained by the vanishing discount approach.

In the context of inventory problems, Iglehart (1963b) and Veinott and Wagner (1965) were the first to study the issue of the existence of an optimal (s, S)-type policy for average cost problems with independent demands, linear holding and backlog costs, and ordering costs consisting of a fixed cost and a proportional variable cost. Iglehart obtained the stationary distribution of the inventory/backlog (or surplus) level given an (s, S) policy using renewal theory arguments (see also Karlin (1958a) and Karlin (1958b)), and developed an explicit formula for the stationary average cost $\mathcal{L}(s, S)$, $s \leq S$, associated with the policy.

Iglehart *assumed* that the function $\mathcal{L}(s, S)$ is continuously differentiable and that there exists a pair (s^*, S^*), $-\infty < s^* < S^* < \infty$, which minimizes $\mathcal{L}(s, S)$ and satisfies the first-order conditions for an interior local minimum. While he does not specify these assumptions explicitly, and certainly does not verify them, he *uses* them in showing the key result that the minimum average cost of a sequence of problems

D. Beyer et al., *Markovian Demand Inventory Models*, International Series in Operations Research & Management Science 108,
DOI 10.1007/978-0-387-71604-6_9,

with increasing finite horizons approaches $\mathcal{L}(s^*, S^*)$ asymptotically as the horizon becomes larger. Veinott and Wagner (1965), with a short additional argument suggested by Derman (1965), were able to advance the Iglehart result into the optimality of the (s^*, S^*) policy in the special case of discrete demands; (see also Veinott (1966)). While Veinott and Wagner deal with the case of discrete demands, they assume, without proving, that Iglehart's results derived for the continuous demand case also hold for their case. It should also be mentioned that Derman's short additional argument also applies to the continuous demand case treated in Iglehart and proves that an $\mathcal{L}(s, S)$-minimizing pair (s^*, S^*), if there exists one, provides an optimal inventory policy; (see Section 9.7).

A good deal of research has been carried out in connection with the average cost (s, S) models since then. With the exception of Zheng (1991) and Huh *et al.* (2008), however, most of this research devoted to establishing the optimality of (s, S) strategies, uses bounds on the inventory position after ordering. Examples are Tijms (1972), Wijngaard (1975), and Küenle and Küenle (1977). On the other hand, quite a few papers are concerned with the computation of the (s^*, S^*) pair that minimizes $\mathcal{L}(s, S)$, and not with the issue of establishing the optimality of an (s, S) policy. Some examples are Stidham (1977), Zheng and Federgruen (1991), Federgruen and Zipkin (1984), Hu *et al.* (1993), and Fu (1994). For other references, the reader is directed to Zheng and Federgruen (1991), Porteus (1985), Sahin (1990), and Presman and Sethi (2006). Furthermore, this literature has generally assumed that together, the papers of Iglehart (1963b) and Veinott and Wagner (1965), have established the optimality of an (s, S)-type policy for the problem.[1] But, this is not quite the case, however, since the assumptions on $\mathcal{L}(s, S)$ *implicit* in Iglehart, to our knowledge, have not been satisfactorily verified.

We will compare the stationary cost analyses of Iglehart (1963b) and Veinott and Wagner (1965) and the vanishing discount approach used in Part II of this book. Both approaches prove the optimality of (s, S) strategies, but for somewhat different notions of the long-run average

[1] Another possible approach views continuous demand distributions as the limit of a sequence of discrete demand distributions. Such a limiting procedure could lead to a proof of optimality of (s, S) policies in the continuous demand case. However, this is by no means a trivial exercise.

cost to be minimized. It turns out that the optimal (s, S) strategies are optimal for these different notions of long-run average cost.

We will reproduce the results of Iglehart (1963b) and Veinott and Wagner (1965) in some detail. Some of the proofs of implicit assumptions are missing in Iglehart (1963b) with respect to the model under his consideration. Without these results, the Iglehart analysis cannot be considered complete. Moreover, these results are by no means trivial.[2]

For this purpose, we need to precisely specify Iglehart's model. Then for this model, we must show that there exists a pair (s^*, S^*), $-\infty < s^* \leq S^* < \infty$ that minimizes $\mathcal{L}(s, S)$. In order to accomplish this, we establish in Section 9.7, *a priori* bounds on the minimizing values of s and S. We should caution that any verification of these assumptions on $\mathcal{L}(s, S)$ should not use arguments that rely on the optimality of an (s, S) policy. With the bounds established in Section 9.7, the continuity of $\mathcal{L}(s, S)$ provides us with the existence of a minimum. Continuous differentiability of $\mathcal{L}(s, S)$ follows from the definition of the surplus cost function $L(y)$ and the assumption of a continuous density for the demand. We then show that if (s^*, S^*) with $s^* = S^*$ is a minimum, then there is another minimum (s, S) with $s < S$. It is then possible to assume that an interior solution always exists, and to obtain it by the first-order conditions of an interior minimum.

Tijms (1972) uses the theory of Markov decision processes (MDP) to prove the optimality of (s, S) strategies for a modified inventory problem with discrete demand. In particular, he imposes upper and lower bounds on the inventory position after ordering. These bounds provide a compact (finite) action space as well as bounded costs. Under these conditions, standard MDP results yield the optimality of an (s, S) strategy.

Zheng (1991) has provided a rigorous proof of the optimality of an (s, S) policy in the case of discrete demands for the model in Veinott and Wagner (1965). He was able to use the theory of countable state Markov decision processes in the case when the solution of the average cost optimality equation for the given problem is bounded, which is clearly not the case here since the inventory cost is unbounded. Note that this theory does not deal with the continuous demand case as it would involve an uncountable state MDP. Zheng relaxed the problem by allowing inventory disposals, and since the inventory costs are charged

[2]In fact, the motivation to write our paper arose from Example 9.4 in Section 9.4. The example shows that even for a well-behaved demand density function satisfying Iglehart's assumptions, the derivation of equation (9.12) crucial for the subsequent analysis requires additional arguments not given in Iglehart's paper. The example also shows that in some cases a base-stock policy can be optimal even in the presence of a positive fixed ordering cost.

on ending inventories in his problem, he obtained a bounded solution for the average cost optimality equation of the relaxed problem that involves a dispose-down-to-S component. But the dispose-down-to-S component of the optimal policy would be invoked in the relaxed problem only in the first period (and only when the initial inventory is larger than S), which has no influence on the long-run average cost of the policy. It follows, therefore, that the (s, S) policy, without the dispose-down-to-S component, will also be optimal for the original problem.

The plan of the chapter is as follows. In Section 9.2, we state the problem under consideration. Section 9.3 summarizes the results of Iglehart relevant to the average cost minimization problem. Furthermore, we point out exactly which implicit assumptions have been used by Iglehart without verification. We develop an example in Section 9.4 to show that even under the quite restrictive assumption of the existence of a continuous demand density, the assumptions implicit in the Iglehart analysis are not necessarily satisfied. In Section 9.5, we derive asymptotic bounds on the minimum cost function. In Section 9.6, we review the analysis contained in Veinott and Wagner (1965) that is devoted to the solution of the average cost problem in the case of discrete demands. To the extent that they use Iglehart's analysis for their solution, we show how their paper is not quite complete and how it can be completed. Section 9.7 contains the proofs needed for the completion of Iglehart's analysis and for establishing the optimality of an (s, S) policy in the continuous demand case. Section 9.8 lists results that connect the stationary distribution approach and the vanishing discount approach; both are undertaken to prove the existence of an optimal (s, S) strategy for the average cost inventory problem. Section 9.9 concludes the chapter.

9.2. Statement of the Problem

In this section we formulate a stationary one-product periodic review inventory model with the following notation and assumptions.

(i) The surplus (inventory/backlog) level at the beginning of period k prior to ordering is denoted by x_k. Unsatisfied demand is fully backlogged.

(ii) The surplus level after ordering, but before demand realizes, in period k is denoted by y_k. Orders arrive immediately.

(iii) The one period demands ξ_k, $k = 1, 2, \ldots$, are i.i.d. and the demand distribution has a density $\varphi(\cdot)$. Let μ denote the mean demand. Assume $0 < \mu < \infty$.

(iv) The surplus cost function $L(y)$, where y is the surplus level immediately after ordering, is given by

$$L(y) = \mathsf{E}(h([y-\xi]^+)) + \mathsf{E}(p([y-\xi]^-)),$$

where $h(\cdot)$ and $p(\cdot)$ represent holding and shortage cost functions, respectively. Furthermore, $L(y)$ is assumed to be convex and finite for all y.

(v) The ordering cost when an amount u is ordered is given by

$$\hat{c}(u) = K\mathbb{1}_{u>0} + cu,\ c \geq 0,\ K > 0.$$

Given an ordering up to policy $Y = (y_1, y_2, \ldots)$, the inventory balance equation is

$$x_{k+1} = y_k - \xi_k.$$

Let $f_n(x|Y)$ denote the expected total cost for an n-period problem with the initial inventory $x_1 = x$ when the order policy Y is used, i.e.,

$$f_n(x|Y) = \mathsf{E}\left\{\sum_{k=1}^{n}[\hat{c}(y_k - x_k) + L(y_k)]\right\}.$$

The objective is to minimize the expected long-run average cost

$$a(x|Y) = \liminf_{n\to\infty} \frac{1}{n} f_n(x|Y) \tag{9.1}$$

over the class of all nonanticipative or history-dependent policies $\mathcal{Y}$. In what follows we use $f_n(x|s,S)$ and $a(x|s,S)$ instead of $f_n(x|Y)$ and $a(x|Y)$, with a slight abuse of notation, if Y is a stationary (s,S) strategy.

The model described above is investigated in a more general setting of Markovian demand and polynomially growing surplus cost in Chapter 6 for the average cost objective function

$$J(x|Y) = \limsup_{n\to\infty} \frac{1}{n} f_n(x|Y). \tag{9.2}$$

There we establish the average cost optimality equation and prove the optimality of an (s,S)-type policy by using the vanishing discount approach.

A policy minimizing either (9.1) or (9.2) does not necessarily minimize the other. However, if an optimal policy Y^* with respect to (9.1) is such that $\lim_{n\to\infty}(1/n)f_n(x|Y^*)$ exists, then this limit is less than or equal to both objective functions associated with any policy $Y \in \mathcal{Y}$. On the other hand, if a policy Y^* is optimal with respect to (9.2) and if $\lim_{n\to\infty}(1/n)f_n(x|Y^*)$ exists, then Y^* may still not minimize (9.1). For this reason, Veinott and Wagner (1965) consider the objective function (9.1) to be the *stronger* of the two; often the term *more conservative* is used instead. In Sections 9.7 and 9.8, we complete Iglehart's stationary state analysis and use Derman's short additional argument to obtain an (s, S) policy, which is optimal with respect to (9.1) and has the limit $(1/n)f_n(x|s, S))$ as $n \to \infty$. Thus, this (s, S) policy also minimizes both objective functions (9.1) and (9.2). In addition, by combining the stationary approach with the dynamic programming approach, we show in Section 9.8 that *any* (s, S) policy that is optimal with respect to (9.2) is also optimal with respect to (9.1).

9.3. Review of Iglehart (1963b)

In this section we will summarize the results of a paper by Iglehart (1963b) relevant to the problem of minimizing the long-run average cost, and point out the implicit assumptions which have been used in his paper without verification.

Let $f_n(x)$ denote the minimal total cost for the n-period problem when the initial surplus level is x, i.e.,

$$f_n(x) = \min_{\mathcal{Y}} f_n(x|Y).$$

The sequence of functions $(f_m(x))_{m=1}^n$ satisfies the dynamic programming equation

$$f_n(x) = \min_{y\geq x}[\hat{c}(y-x) + L(y) + \int_0^\infty f_{n-1}(y-\xi)\varphi(\xi)d\xi]. \tag{9.3}$$

Furthermore, it is known that $f_m(x)$ is K-convex, and an optimal strategy minimizing the total cost is determined by a sequence $(s_m, S_m)_{m=1}^n$ of real numbers with $s_m \leq S_m$, such that the optimal order quantity in period m is

$$u_m = \begin{cases} S_m - x_m & \text{if} \quad x_m \leq s_m, \\ 0 & \text{if} \quad x_m > s_m, \end{cases}$$

where x_m denotes the surplus level at the beginning of the mth period.[3]

It is obvious that $\lim_{n\to\infty} f_n(x) = \infty$ for all initial surplus levels x. Iglehart investigates the asymptotic behavior of the function $f_n(x)$ for large n. Heuristic arguments suggest that for a stationary infinite hori-

[3]Well-known papers dealing with this finite horizon problem are those of Scarf (1960), Schäl (1976), and Veinott (1966).

zon inventory problem, a stationary strategy should be optimal. Furthermore, it is reasonable to expect that this stationary strategy is of (s, S)-type.

Iglehart obtains the stationary distribution of the surplus level and the expected one-period cost under any given (s, S) strategy satisfying $-\infty < s \leq S < \infty$. He uses the result of Karlin (1958a,b) that the surplus level x_n at the beginning of period n converges in distribution to a random variable whose distribution has the density

$$f(x) = \begin{cases} \dfrac{m(S-x)}{1+M(\Delta)}, & s < x \leq S, \\ \dfrac{h(\Delta, s-x)}{1+M(\Delta)}, & x \leq s, \end{cases} \tag{9.4}$$

where $\Delta := S - s$, $M(\cdot)$ and $m(\cdot)$ are the renewal function and the renewal density associated with $\varphi(\cdot)$, respectively, and $h(\Delta, \cdot)$ is the density of the order quantity in excess of Δ. Note that by the Elementary Renewal Theorem B.3.1, we have $M(t)/t \to 1/\mu$ as $t \to \infty$. Also $M(t) \to \infty$ as $t \to \infty$.

Furthermore, the one-period cost $C(x)$, given the initial surplus x, is

$$C(x) = \begin{cases} K + L(S) + c(S-x), & x \leq s, \\ L(x), & s < x \leq S. \end{cases} \tag{9.5}$$

Averaging $C(x)$ with respect to $f(x)$ yields the following formula for the stationary cost $\mathcal{L}(s, S)$ per period corresponding to the strategy parameters s and S:

$$\begin{aligned} \mathcal{L}(s,S) &= \int_{-\infty}^{s} \Big[K + L(S) + c(S-x)\Big] f(x)dx + \int_{s}^{S} L(x)f(x)dx \\ &= \frac{K + L(S) + \int_s^S L(x)m(S-x)dx}{1+M(S-s)} + c\mu. \end{aligned} \tag{9.6}$$

Remark 9.1 It follows from the convergence in distribution of the surplus level in period n that the expected cost $\mathsf{E}(C(x_n))$ in period n converges to $\mathcal{L}(s, S)$. Therefore, we have

$$a(x|s,S) = \lim_{n\to\infty} \frac{1}{n} f_n(x|s,S) = \lim_{n\to\infty} \frac{1}{n}\left(\sum_{i=1}^{n} \mathsf{E}(C(x_n))\right) = \mathcal{L}(s,S).$$

In many cases, it is more convenient to define the stationary cost in terms of Δ and S:

$$\tilde{\mathcal{L}}(\Delta, S) := \mathcal{L}(S-\Delta, S) = \frac{K + L(S) + \int_0^\Delta L(S-x)m(x)dx}{1+M(\Delta)} + c\mu, \quad (9.7)$$

or in terms of s and Δ:

$$\begin{aligned}\hat{\mathcal{L}}(s, \Delta) &:= \mathcal{L}(s, s+\Delta) \\ &= \frac{K + L(s+\Delta) + \int_0^\Delta L(s+\Delta-x)m(x)dx}{1+M(\Delta)} + c\mu. \quad (9.8)\end{aligned}$$

Iglehart then attempts to minimize $\mathcal{L}(s, S)$ with respect to s and S. Implicitly, he assumes $\mathcal{L}(s, S)$ to be continuously differentiable and that the minimum is attained for some s^* and S^* satisfying $-\infty < s^* < S^* < \infty$, i.e., the minimum is attained at an interior point and not at the boundary $s = S$ of the feasible parameter set. Therefore, he can take derivatives of $\hat{\mathcal{L}}(s, \Delta)$ with respect to s and Δ and obtain necessary conditions for the minimum by setting them equal to zero:

$$\frac{\partial \hat{\mathcal{L}}(s, \Delta)}{\partial s} = L'(s+\Delta) + \int_0^\Delta L'(s+\Delta-x)m(x)dx = 0 \quad (9.9)$$

and

$$\begin{aligned}&\frac{\partial \hat{\mathcal{L}}(s, \Delta)}{\partial \Delta} \\ &= \frac{L'(s+\Delta) + \int_0^\Delta L'(s+\Delta-x)m(x)dx + L(s)m(\Delta)}{1+M(\Delta)} \\ &\quad - \frac{K + L(s+\Delta) + \int_0^\Delta L(s+\Delta-x)m(x)dx}{(1+M(\Delta))^2} m(\Delta) = 0. \quad (9.10)\end{aligned}$$

Note that if $s^* = S^*$, Condition (9.9) must be relaxed to $\partial \hat{\mathcal{L}}/\partial s \leq 0$. Combining the necessary Conditions (9.9) and (9.10), one obtains

$$(1+M(\Delta))L(s) - K - L(s+\Delta) - \int_0^\Delta L(s+\Delta-x)m(x)dx)m(\Delta) = 0. \quad (9.11)$$

Assuming further that $m(\Delta^*) = m(S^* - s^*) > 0$ for the minimizing values s^* and S^*, Iglehart obtains the following formula, crucial for his subsequent analysis, by dividing by $m(\Delta)$:

$$L(s^*) = \frac{K + L(s^* + \Delta^*) + \int_0^{\Delta^*} L(s^* + \Delta^* - x)m(x)dx}{1 + M(\Delta^*)} = \mathcal{L}(s^*, S^*) - c\mu. \tag{9.12}$$

Let us recapitulate Iglehart's implicit assumptions here. For his subsequent analysis, he requires that

(I1) there is a pair (s^*, S^*) with $-\infty < s^* < S^* < \infty$ (i.e., an interior solution) that minimizes $\mathcal{L}(s, S)$,

(I2) $\mathcal{L}(s, S)$ is continuously differentiable, and

(I3) the minimizing pair (s^*, S^*) satisfies $m(S^* - s^*) > 0$.

We will see that there is always a pair (s^*, S^*) satisfying Assumption (I1) since $K > 0$. On the other hand, in general, not every minimizer of $\mathcal{L}(s, S)$ satisfies Assumption (I1) even though $K > 0$; (see Example 9.4). Assumption (I2) is satisfied because of the continuous differentiability of $L(\cdot)$ and the existence of a continuous density, which implies continuous differentiability of $M(\cdot)$. Assumption (I3) is more difficult to deal with. In Example 9.4, Assumption (I3), i.e., $m(S^* - s^*) > 0$, is violated for all minimizing (s^*, S^*) pairs. The fact that Assumption (I3) does not hold in general is a problem that can be fixed. It turns out that it is not Assumption (I3) itself but rather equation (9.12) derived with the help of this assumption that is crucial for the subsequent analysis. In Section 9.7, we prove that there is always a minimizing pair (s^*, S^*) that satisfies (9.12), even if $m(S^* - s^*) = 0$ for all minimizing pairs.

9.4. An Example

We develop an example with $K > 0$, in which $s = S$ is optimal and $m(S^* - s^*) = 0$ for all minimizing pairs of parameters (s^*, S^*). To do so, we prove two preliminary results.

LEMMA 9.1 *Let* $L(\cdot)$ *be a convex function with* $\lim_{x \to \pm\infty} L(x) = \infty$. *Then for* $0 \leq x_1 \leq x_2$,

$$\min_S \{L(S) + L(S - x_1)\} \leq \min_S \{L(S) + L(S - x_2)\}.$$

Proof. Because of the limit property and the convexity of $L(\cdot)$, $L(S) + L(S - x_2)$ attains its minimum. Let S_2 be a minimum point of $L(S) +$

$L(S-x_2)$. Let us first assume that $L(S_2) \geq L(S_2-x_2)$. Then, it follows from the convexity of $L(\cdot)$ that $L(S_2-x_2+x_1) \leq L(S_2)$, and we obtain

$$\begin{aligned} \min_S\{L(S)+L(S-x_1)\} &\leq L(S_2-x_2+x_1)+L(S_2-x_2)\} \\ &\leq L(S_2)-L(S_2-x_2) \\ &= \min_S\{L(S)+L(S-x_2)\}. \end{aligned}$$

If $L(S_2) < L(S_2-x_2)$, we conclude from the convexity of $L(\cdot)$ that $L(S_2-x_1) \leq L(S_2-x_2)$. Thus,

$$\begin{aligned} \min_S\{L(S)+L(S-x_1)\} &\leq L(S_2)+L(S_2-x_1)\} \\ &\leq L(S_2)+L(S_2-x_2) \\ &= \min_S\{L(S)+L(S-x_2)\}, \end{aligned}$$

which completes the proof. □

LEMMA 9.2 *Let $L(\cdot)$ be a convex function with $\lim_{x\to\pm\infty} L(x) = \infty$. Then*

$$\lim_{D\to\infty} \min_S\{L(S)+L(S-D)\} = \infty.$$

Proof. Fix $D > 0$ and let S^* denote a fixed minimum point of $L(\cdot)$. It is easy to see that

$$L(S) \geq \min\{L(S^*+D/2), L(S^*-D/2)\} \text{ for } |S-S^*| \geq D/2.$$

Also, substituting $S-D$ for S,

$$L(S-D) \geq \min\{L(S^*+D/2), L(S^*-D/2)\} \text{ for } |S-D-S^*| \geq D/2.$$

Since at least one of the above conditions is satisfied and $L(x) \geq L(S^*)$ for all x, we obtain

$$L(S)+L(S-D) \geq \min\{L(S^*+D/2), L(S^*-D/2)\} + L(S^*).$$

Therefore, it follows that

$$\begin{aligned} &\lim_{D\to\infty} \min_S\{L(S)+L(S-D)\} \\ &\geq \lim_{D\to\infty} \min\{L(S^*+D/2), L(S^*-D/2)\} + L(S^*) = \infty, \end{aligned}$$

and the proof is completed. □

For the purpose of this example, we assume the unit purchase cost $c = 0$. It can easily be extended to the case $c > 0$. Let the one-period demand consist of a deterministic component $D \geq 0$ and a random component d. Let $\varphi(\cdot)$ be the density of d, where $\varphi(\cdot)$ is continuous on $(-\infty, \infty)$, $\varphi(t) = 0$ for $t \leq 0$, and $\varphi(t) > 0$ for $t \in (0, \varepsilon)$ for some $\varepsilon > 0$. The density of the one-period demand is then given by $\varphi^D(t) = \varphi(t-D)$. It is continuous on the entire real line. We denote the renewal density and the renewal function with respect to φ^D by m^D and M^D, respectively. Let

$$L^D(x) = \int_{-\infty}^{\infty} l(x-\xi)\varphi^D(\xi)d\xi$$

be the expected one-period surplus cost function. Let $L(\cdot) = L^0(\cdot)$. Then, we have

$$L^D(x) = L(x-D).$$

Now the stationary cost function, given s and S, is

$$\mathcal{L}^D(s,S) = \frac{K + L^D(S) + \int_s^S L^D(x)m^D(S-x)dx}{1 + M^D(S-s)},$$

and in view of $\Delta = S - s$, we define

$$\tilde{\mathcal{L}}^D(\Delta, S) = \mathcal{L}^D(S-\Delta, S) = \frac{K + L^D(S) + \int_0^\Delta L^D(S-x)m^D(x)dx}{1 + M^D(\Delta)}.$$

We will show that there is a constant D_0 such that for all $D > D_0$, the pair that minimizes the stationary average cost is of the following form

> (s^*, S^*) is a minimum point of $\mathcal{L}^D(s,S)$ if, and only if, S^* minimizes $L^D(S)$ and $0 \leq S^* - s^* \leq D$.

It is obvious that because of $\varphi^D(t) = 0$ for all $t \leq D$, $m^D(t) = 0$ and $M^D(t) = 0$ for $t \leq D$. Therefore, for any given $\Delta \leq D$,

$$\min_S \tilde{\mathcal{L}}^D(\Delta, S) = K + \min_S L^D(S) = K + \min_S L(S).$$

We will show that for a sufficiently large D,

$$\min_S \tilde{\mathcal{L}}^D(\Delta, S) > K + \min_S L(S) \text{ for all } \Delta > D.$$

To do so, we consider three cases that arise when $\Delta > D$ for any given D.

- Case 1. Let $\Delta \in \{\Delta : 0 < M^D(\Delta) \le 1\}$. Then,

$$
\begin{aligned}
&\min_S \tilde{\mathcal{L}}^D(\Delta, S) \\
&= \frac{K + \min_S \left\{ L^D(S) + \int_0^\Delta L^D(S-x) m^D(x) dx \right\}}{1 + M^D(\Delta)} \\
&\ge \frac{K + \min_S \left\{ L^D(S) + \min_{x \in [D,\Delta]} \{L^D(S-x)\} M^D(\Delta) \right\}}{1 + M^D(\Delta)} \\
&= K + \frac{1}{1 + M^D(\Delta)} \left(\min_S \left\{ (1 - M^D(\Delta)) L^D(S) \right.\right. \\
&\quad \left.\left. + \min_{x \in [D,\Delta]} \{L^D(S-x) + L^D(S) - K\} M^D(\Delta) \} \right\} \right) \\
&\ge K + \frac{1}{1 + M^D(\Delta)} \left((1 - M^D(\Delta)) \min_S L^D(S) \right. \\
&\quad \left. + M^D(\Delta) \min_S \{ \min_{x \in [D,\Delta]} \{L^D(S-x) + L^D(S) - K\}\} \right).
\end{aligned}
\tag{9.13}
$$

Using Lemma 9.1, we have

$$
\begin{aligned}
&\min_S \{ \min_{x \in [D,\Delta]} \{L^D(S-x) + L^D(S) - K\}\} \\
&\quad = \min_S \{ \min_{x \in [D,\Delta]} \{L(S-D-x) + L(S-D) - K\}\} \\
&\quad = \min_S \{ \min_{x \in [D,\Delta]} \{L(S-x) + L(S) - K\}\} \\
&\quad = \min_S \{L(S-D) + L(S) - K\}.
\end{aligned}
\tag{9.14}
$$

On account of Lemma 9.2, this expression tends to infinity as $D \to \infty$. Therefore, there is a $D_1 > 0$ such that

$$\min_S \{L(S-D) + L(S) - K\} > 2 \min_S L(S) \text{ for all } D \ge D_1. \tag{9.15}$$

Thus from (9.13)–(9.15), we have

$$\min_S \tilde{\mathcal{L}}^D(\Delta, S) > K + \min_S L(S) \text{ for all } D \ge D_1,$$

implying that Δ in this case cannot be a minimizer.

- Case 2. Let $\Delta \in \{\Delta : 1 < M^D(\Delta) \le 2\}$. Then,

$$\begin{aligned}
\tilde{\mathcal{L}}^D(\Delta, S) &= \frac{K + L^D(S) + \int_0^\Delta L^D(S-x) m^D(x) dx}{1 + M^D(\Delta)} \\
&\ge \frac{K + \frac{1}{2}\int_0^\Delta (L^D(S-x) + L^D(S)) m^D(x) dx}{3} \\
&\ge \frac{1}{6} \min_{x \in [D,\Delta]} \{L^D(S-x) + L^D(S)\} M^D(\Delta) \\
&\ge \frac{1}{6} \min_{x \in [D,\Delta]} \{L^D(S-x) + L^D(S)\}.
\end{aligned}$$

In view of Lemma 9.1, taking the minimum with respect to S yields

$$\begin{aligned}
\min_S \tilde{\mathcal{L}}^D(\Delta, S) &\ge \frac{1}{6} \min_S \{ \min_{x \in [D,\Delta]} \{L^D(S-x) + L^D(S)\}\} \\
&= \frac{1}{6} \min_S \{L^D(S-D) + L^D(S)\} \\
&= \frac{1}{6} \min_S \{L(S-D) + L(S)\}. \qquad (9.16)
\end{aligned}$$

From Lemma 9.2 we know that this expression tends to infinity as $D \to \infty$. Therefore, there is a $D_2 > 0$ such that

$$\min_S \{L(S-D) + L(S)\} > 6(K + \min_S L(S)) \text{ for all } D \ge D_2. \quad (9.17)$$

Then, from (9.16) and (9.17),

$$\min_S \tilde{\mathcal{L}}^D(\Delta, S) > K + \min_S L(S) \text{ for all } D \ge D_2,$$

and thus Δ in Case 2 cannot be optimal.

- Case 3. Let $\Delta \in \{\Delta : 2 < M^D(\Delta)\}$. Now we define $C := 3(K + \min_S L(S)) = 3(K + \min_S L^D(S))$, $S_l := \min\{S : L^D(S) \le C\}$, and $S_u := \max\{S : L^D(S) \le C\}$. It follows from the convexity of $L^D(\cdot)$ and the fact that $\lim_{x \to \pm\infty} L^D(x) = \infty$ that S_l and S_u are finite. Let $D_0 := \max\{D_1, D_2, S_u - S_l\}$ and choose $D \ge D_0$. Then we obtain

$$\begin{aligned}
\tilde{\mathcal{L}}^D(\Delta, S) &= \frac{K + L^D(S) + \int_0^\Delta L^D(S-x) m^D(x) dx}{1 + M^D(\Delta)} \\
&\ge \frac{\int_0^\Delta L^D(S-x) m^D(x) dx}{1 + M^D(\Delta)}.
\end{aligned}$$

It is easy to conclude from the definitions of S_l and S_u that $L^D(S - x) \geq C$ for $x \in [0, \Delta] \setminus [S - S_u, S - S_l]$. Because

$$[0, \Delta] \setminus [S - S_u, S - S_l] = [0, \Delta] \setminus [(S - S_u)^+, (S - S_l)^+],$$

we obtain

$$\tilde{\mathcal{L}}^D(\Delta, S) \geq \frac{CM(\Delta) - C(M^D((S - S_l)^+) - M^D((S - S_u)^+))}{1 + M^D(\Delta)}. \tag{9.18}$$

Since $0 \leq (S - S_l)^+ - (S - S_u)^+ \leq S_u - S_l \leq D$, it is clear that there is almost surely at most one renewal between $(S - S_u)^+$ and $(S - S_l)^+$, and therefore $M^D((S - S_l)^+) - M^D((S - S_u)^+) \leq 1$. Thus,

$$\tilde{\mathcal{L}}^D(\Delta, S) \geq \frac{C(M^D(\Delta) - 1)}{1 + M^D(\delta)} > \frac{C}{3} = K + \min_S L(S).$$

Therefore, for $D > D_0$ and all $\Delta = S - s > D$, we have

$$\min_S \mathcal{L}^D(s, S) = \min_S \tilde{\mathcal{L}}^D(\Delta, S) \geq \min_S \mathcal{L}^D(S, S).$$

Because $\mathcal{L}^D(s, S)$ is constant in s for $S - D \leq s \leq S$, it is clear that the minimizing point (s^*, S^*) satisfies the desired conditions. Furthermore, because $m^D(t) = 0$ for $0 \leq t \leq D$, it holds that

$$m^D(S^* - s^*) = 0 \text{ for all minimum points of } \mathcal{L}^D(s, S).$$

9.5. Asymptotic Bounds on the Optimal Cost Function

As the horizon n of the inventory problem becomes large, it is reasonable to expect that the optimal strategy parameters (s_m, S_m) and (s_{m+1}, S_{m+1}) for small m do not differ significantly and that the optimal strategy tends to a stationary one. On the other hand, if a stationary strategy is applied, the inventory level tends towards a steady state. The minimum cost per period that one can achieve in the steady state is $k := \mathcal{L}(s^*, S^*)$. If the system approaches the steady state fast enough, one could expect the difference $f_n(x) - nk$ to be uniformly bounded with respect to n for any x. In Section 4 of his paper, Iglehart obtains bounds on $f_n(x)$ in terms of an explicitly given solution $\psi(\cdot)$ of the equation

$$\psi(x) = \min_{y \geq x}[\hat{c}(y - x) + L(y) - k + \int_0^\infty \psi(y - \xi)\varphi(\xi)d\xi]. \tag{9.19}$$

He proves that the function $\psi(\cdot)$ defined as

$$\psi(s^* + y) = \begin{cases} -cy, & y \leq 0, \\ L(y + s^*) - k + \int_0^\infty \psi(y + s^* - \xi)\varphi(\xi)d\xi, & y > 0, \end{cases} \tag{9.20}$$

satisfies (9.19). The pair (s^*, S^*) is a minimizer of $\mathcal{L}(s, S)$, which satisfies (9.12).

Briefly, the proof goes as follows. First, Iglehart verifies that the function $\psi(\cdot)$ defined in (9.20) is K-convex.

Remark 9.2 It should be mentioned that $\psi(\cdot)$ is also K-convex if we replace k by any value larger than $\mathcal{L}(s^*, S^*) = L(s^*) + c\mu$.

In the next step, he shows that the function

$$G(y) = cy + L(y) + \int_0^\infty \psi(y - \xi)\varphi(\xi)d\xi,$$

which represents the function to be minimized in (9.19), attains its minimum at $y = S^*$, and $G(s^*) = K + G(S^*)$ for $\psi(\cdot)$ defined in (9.20). To show this, it is essential that (9.12) holds.

The K-convexity of $G(\cdot)$ follows from the K-convexity of $\psi(\cdot)$. Now we return to equation (9.19) written in terms of $G(\cdot)$, i.e.,

$$\psi(x) = -k - cx + \min_{y \geq x}[K\mathbb{1}_{y>x} + G(y)],$$

and transform its RHS. For $x \leq s^*$, the minimum is attained for $y = S^*$ and we get

$$\begin{aligned} -k - cx + \min_{y \geq x}&[K\mathbb{1}_{y>x} + G(y)] \\ &= -k - cx + K + G(S^*) = -k - cx + G(s^*) \\ &= -k - cx + cs^* + L(s^*) + \int_0^\infty \psi(s^* - \xi)\varphi(\xi)d\xi \\ &= -c(x - s^*). \end{aligned}$$

For $x > s^*$, the minimum is attained for $y = x$ and we obtain

$$\begin{aligned} -k - cx + \min_{y \geq x}&[K\mathbb{1}_{y>x} + G(y)] \\ &= -k - cx + G(x) \\ &= -k + L(x) + \int_0^\infty \psi(x - \xi)\varphi(\xi)d\xi. \end{aligned}$$

Therefore, the function $\psi(\cdot)$ defined in (9.20) actually satisfies (9.19). Observe that we can write explicitly,

$$G(y+s^*) = \begin{cases} c(s^*+\mu) + L(y+s^*) & \text{for} \quad y < 0, \\ c(s^*+\mu) + L(s^*) + L(y+s^*) & \\ \qquad + \int_0^y L(y+s^*-\xi)m(\xi)d\xi & \\ \qquad -L(s^*)[1+M(y)] & \text{for} \quad y \geq 0. \end{cases}$$

The main result of this section is that for given $W \in \mathbb{R}$, there are constants r and R depending on W such that the inequalities

$$nk + \psi(x) - r \leq f_n(x) \leq nk + \psi(x) + R, \text{ for } x \leq W, \tag{9.21}$$

hold. The assertion is proved by induction. First, Iglehart shows that the optimal order levels S_k for the n-period problem are uniformly bounded from above, i.e., the bound does not depend on n. Then he chooses a constant W larger than this bound and proves that the inequality holds for $n = 1$. Since $f_1(x)$ and $\psi(x)$ are both linear with slope $-c$ for $x \leq \min\{s_1, s\}$, we can set

$$r = \min_{\min\{s_1,s\} \leq x \leq W} \{f_1(x) - \psi(x) - k\}$$

and

$$R = \max_{\min\{s_1,s\} \leq x \leq W} \{f_1(x) - \psi(x) - k\}.$$

The induction step for $n = N+1$ uses (9.3) and (9.19). Note that because $x \leq W$ and $S_n \leq W$, the minimum in (9.3) is attained for some $y \leq W$, and the inequality (9.21) can be used for $n = N$.

Because $\psi(\cdot)$ is continuous and therefore $\psi(x) < \infty$ for any x, it follows from (9.21) that

$$\lim_{n\to\infty} \frac{f_n(x)}{n} = k. \tag{9.22}$$

It should be mentioned that the proof of (9.21) requires only that $\psi(\cdot)$ solves (9.19) and that $\psi(x)$ is linear with slope $-c$ for all x smaller than some finite constant. Moreover, it follows from (9.22) that there is no such solution for $k \neq L(s^*) + c\mu$.

In Sections 6 and 7, Iglehart investigates the limiting behavior of the function $g_n(x) = f_n(x) - nk$ as $n \to \infty$. He proves that for $K = 0$,

$$\lim_{n\to\infty} g_n(x) = \psi(x) + A,$$

where $\psi(\cdot)$ is given by (9.20) and A is a constant. For $K > 0$, he is only able to obtain

$$\limsup_{n\to\infty} g_n(x) \leq \psi(x) + B$$

for a constant B. For $K > 0$, he also conjectures that $\liminf_{n\to\infty} g_n(x) \geq \psi(x) + B$, which would lead to $\lim_{n\to\infty} g_n(x) = \psi(x) + B$.

9.6. Review of the Veinott and Wagner Paper

Veinott and Wagner (1965) deal with the inventory problem introduced in Section 9.2 with one essential difference. They consider the demand ξ_i in period i to be a *discrete* random variable taking nonnegative integer values. They assume that one-period demands $\xi_1, \xi_2, \ldots$ are i.i.d. random variables with the probability

$$\mathsf{P}(\xi_i = k) = \varphi(k), \qquad k = 0, 1, \ldots, \quad i = 1, 2, \ldots.$$

We only recapitulate here the results of Veinott and Wagner that are important in the context of the existence of an optimal (s, S) strategy. We restrict our attention to the case of the zero leadtime for convenience in exposition.

The discrete renewal density and the renewal function are defined as

$$\begin{aligned} m(k) &= \sum_{i=1}^{\infty} \varphi^i(k), \\ M(k) &= \sum_{i=1}^{\infty} \Phi^i(k) = \sum_{j=0}^{k} m(j), \qquad k = 0, 1, \ldots, \end{aligned}$$

where φ^i and Φ^i denote the probabilities and the cumulative distribution function of the i-fold convolution of the demand distribution, respectively.

Employing a renewal approach or a stationary probability approach, Veinott and Wagner derive a formula for the stationary average cost $a(x|s, S)$, given a particular stationary (s, S) strategy. Since the unit purchase cost does not influence the optimal strategy, the formulas are derived for the case $c = 0$. An extension to $c > 0$ is straightforward. Veinott and Wagner (VW) obtain

$$a(x|s, S) = \mathcal{L}_{\mathrm{VW}}(s, S) = \frac{K + L(S) + \sum_{i=0}^{S-s} L(S - j)m(j)}{1 + M(S - s)}$$

for integer values of s and S. It should be mentioned that this function does not depend on the initial surplus x.

For their discrete demand case, Veinott and Wagner claim that, just as in Iglehart's continuous demand density case, a minimizing pair (s^*, S^*)

of $\mathcal{L}_{\text{VW}}(s, S)$ would satisfy (9.22), i.e.,

$$\lim_{n \to \infty} \frac{1}{n} f_n(x) = \mathcal{L}_{\text{VW}}(s^*, S^*). \tag{9.23}$$

Furthermore, in the appendix of their paper, Veinott and Wagner establish the bounds for the parameters minimizing $\mathcal{L}_{\text{VW}}(s, S)$. The proofs of these bounds, derived in their paper for computational purposes, do not require the existence of a stationary optimal (s, S) strategy in the average cost case, but critically depend on the discrete nature of the demand.

Having established bounds on the minimizers of $\mathcal{L}_{\text{VW}}(s, S)$, it is clear that the discrete function $\mathcal{L}_{\text{VW}}(s, S)$ attains its minimum for an integer pair (s^*, S^*).

However, the claim by Veinott and Wagner that a discrete analog of Iglehart's analysis yields (9.23) for the discrete demand case, requires some additional arguments not included in their paper. Observe that a completion of Iglehart's analysis not only requires the existence of a finite minimizer of the stationary average cost function, but it also needs the minimizer to satisfy equation (9.12), which with $c = 0$ reduces to

$$\mathcal{L}(s^*, S^*) = L(s^*). \tag{9.24}$$

In general, the integer minimizer of $\mathcal{L}_{\text{VW}}(s, S)$ will not satisfy this condition.

Subsequently, in order to establish the dynamic programming equation in the MDP context, Tijms (1972) has shown that there are integer minimizers (s^*, S^*) of $\mathcal{L}(s, S)$ such that

$$L(s^* - 1) \geq \mathcal{L}(s^*, S^*) \geq L(s^*). \tag{9.25}$$

For our purpose, it immediately follows from (9.25) and the continuity of $L(x)$ that there is an $s^{\#} < S^*$ such that

(a) $(\lfloor s^{\#} \rfloor, S^*)$ is an integer minimizer of $\mathcal{L}_{\text{VW}}(s, S)$, implying that $\mathcal{L}_{\text{VW}}(\lfloor s^{\#} \rfloor, S^*) = \mathcal{L}_{\text{VW}}(s^*, S^*)$, and

(b) $\mathcal{L}_{\text{VW}}(\lfloor s^{\#} \rfloor, S^*) = L(s^{\#})$,

where $\lfloor x \rfloor$ denotes the largest integer smaller or equal to x. Using the pair $(s^{\#}, S^*)$ and the function $\mathcal{L}(s, S) := \mathcal{L}_{\text{VW}}(\lfloor s \rfloor, S)$ in Iglehart's analysis, we can get the desired formula (9.24).

Once (9.24) is established, the short additional argument suggested by Derman and used by Veinott and Wagner would provide the optimality of a stationary (s, S) strategy for the average cost inventory problem.

Specifically, it follows from the definition of f_n that $f_n(x) \leq f_n(x|Y)$ for all initial values x and all history-dependent strategies $Y \in \mathcal{Y}$. Thus,

$$a(x|s^*, S^*) \quad = \quad \mathcal{L}(s^*, S^*) = \lim_{n \to \infty} \frac{1}{n} f_n(x)$$

$$= \liminf_{n\to\infty} \frac{1}{n} f_n(x) \le \liminf_{n\to\infty} \frac{1}{n} f_n(x|Y) = a(x|Y),$$

for any history-dependent strategy $Y \in \mathcal{Y}$. Therefore, (s^*, S^*) is average optimal.

We now return to the continuous demand case of Iglehart.

9.7. Existence of Minimizing Values of s and S

In this section, we first establish *a priori* bounds on the values of s and S that minimize $\mathcal{L}(s, S)$. Once that is done, the continuity of $\mathcal{L}(s, S)$ ensures the existence of a solution (s^*, S^*) that minimizes $\mathcal{L}(s, S)$. While Veinott and Wagner have proved bounds on the minimizing s and S, their proofs use the fact that the demands are discrete, and there is no obvious way to transfer their proofs to the continuous demand case. Additionally, they employ a vanishing discount argument. Since the discounted problem is not within the scope of the original Iglehart paper, we will provide bounds on the minimizing parameters (s, S) using only the properties of the stationary cost function.

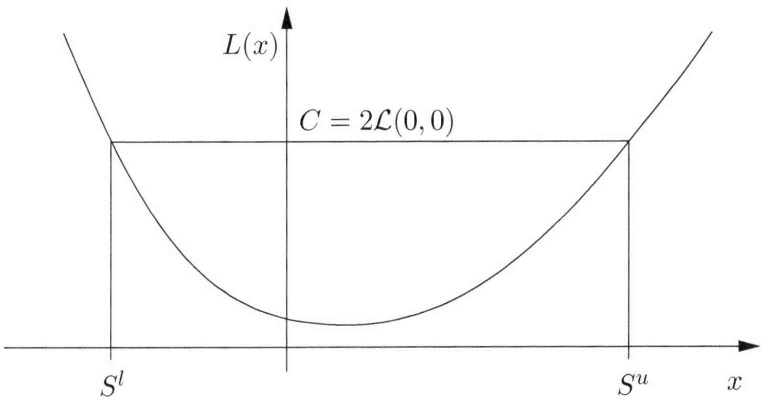

Figure 9.1. Definitions of S^l and S^u

LEMMA 9.3 *There is a constant $\bar{S}$ such that for all $S \ge \bar{S}$ and $s \le S$, the stationary cost $\mathcal{L}(s, S) > \mathcal{L}(0, 0)$.*

Proof. Let $C = 2\mathcal{L}(0,0) > 0$. Choose S^l and S^u such that $S^l \le S^u$ and $L(S^l) = L(S^u) = C$; (see Figure 9.1). Let $S \ge S^u$. Our proof requires three cases to be considered: $s \le S^l$, $S^l < s < S^u$, and $S^u \le s \le S$.

- Case $s \le S^l$. In this case, $M(S - S^l) \le M(S - s)$, which we will use later. From the cost formula (9.5) and the stationary probability density (9.4), we have

$$\begin{aligned}
\mathcal{L}(s,S) &\geq \int_s^S L(x)f(x)dx \geq C\int_s^{S^l} f(x)dx + C\int_{S^u}^S f(x)dx \\
&= \frac{C}{1+M(S-s)}\left[\int_s^S m(S-x)dx - \int_{S^l}^{S^u} m(S-x)dx\right] \\
&= \frac{C}{1+M(S-s)}\left[M(S-s) - M(S-S^l) + M(S-S^u)\right] \\
&= C - C\left[\frac{1+M(S-S^l)-M(S-S^u)}{1+M(S-s)}\right] \\
&\geq C - C\left[\frac{1+M(S-S^l)-M(S-S^u)}{1+M(S-S^l)}\right] \\
&= C\left[\frac{M(S-S^u)}{1+M(S-S^l)}\right]. \qquad (9.26)
\end{aligned}$$

- Case $S^l < s < S^u$. In this second case, we use (9.5) and (9.4) to immediately obtain

$$\begin{aligned}
\mathcal{L}(s,S) \geq \int_{S^u}^S L(x)f(x)dx &= \frac{C}{1+M(S-s)}\int_{S^u}^S m(S-x)dx \\
&= C\left[\frac{M(S-S^u)}{1+M(S-s)}\right].
\end{aligned}$$

But in this case, $S - s \leq S - S^l$. Therefore,

$$\mathcal{L}(s,S) \geq C\left[\frac{M(S-S^u)}{1+M(S-S^l)}\right]. \qquad (9.27)$$

Since $M(t)/t \to 1/\mu$ as $t \to \infty$, the expression in the square brackets in (9.26) and (9.27), which does not depend on s, goes to one as $S \to \infty$, i.e.,

$$\begin{aligned}
\lim_{S\to\infty} \frac{M(S-S^u)}{1+M(S-S^l)} &= \lim_{S\to\infty} \frac{M(S-S^u)}{M(S-S^l)} \\
&= \lim_{S\to\infty} \frac{M(S-S^u)}{S-S^u}\frac{S-S^l}{M(S-S^l)}\frac{S-S^u}{S-S^l} \\
&= \frac{1}{\mu}\cdot\mu\cdot 1 = 1.
\end{aligned}$$

Therefore, there is an $\bar{S} \geq S^u$, independent of s, such that $M(S-S^u)/(1+M(S-S^l)) > 1/2$ for all $S \geq \bar{S}$. This means that

$$\mathcal{L}(s,S) > \mathcal{L}(0,0) \text{ for all } S \geq \bar{S} \text{ and } s < S^u,$$

which proves the lemma in the first two cases.

- Case $S^u \leq s \leq S$. Finally, for this third case, we use (9.5) to obtain

$$\mathcal{L}(s,S) = \int_{-\infty}^{s} [K + L(S)]f(x)dx + \int_{s}^{S} L(x)f(x)dx \geq C > \mathcal{L}(0,0).$$

This completes the proof. □

LEMMA 9.4 *For S^l defined in Lemma 9.3, $\mathcal{L}(s,S) > \mathcal{L}(0,0)$ for all $S \leq S^l$ and $s \leq S$.*

Proof. For $S \leq S^l$, it is clear from (9.5) that

$$\mathcal{L}(s,S) = \int_{-\infty}^{s} [K + L(S)]f(x)dx + \int_{s}^{S} L(x)f(x)dx \geq C > \mathcal{L}(0,0).$$

This completes the proof. □

It is easy to see that together, Lemmas 9.3 and 9.4, prove that the minimizing value of S lies in the set $[S^l, \bar{S}]$. In the next lemma, we show that the minimizing value of s is bounded as well.

LEMMA 9.5 *There is a constant $\bar{s}$ such that for all $s \leq \bar{s}$ and $S \geq s$, $\mathcal{L}(s,S) > \mathcal{L}(0,0)$.*

Proof. Let S^l and $\bar{S}$ be defined as in Lemma 9.3. Let $s \leq S^l$. For any S satisfying $s \leq S \leq S^l$ or $s \leq S^l \leq \bar{S} \leq S$, it follows from Lemmas 9.3 and 9.4 that $\mathcal{L}(s,S) > \mathcal{L}(0,0)$. Therefore, we can restrict our attention to the values of S which satisfy $S^l \leq S \leq \bar{S}$. Then, from (9.5) and (9.4) we have

$$\begin{aligned}
\mathcal{L}(s,S) &\geq \int_{s}^{S^l} L(x)f(x)dx \\
&\geq C\int_{s}^{S^l} f(x)dx = \frac{C}{1+M(S-s)}\left[\int_{s}^{S^l} m(S-x)dx\right]
\end{aligned}$$

$$
\begin{aligned}
&= \frac{C}{1+M(S-s)}\left[M(S-s)-M(S-S^l)\right] \\
&= C - C\left[\frac{1+M(S-S^l)}{1+M(S-s)}\right] \\
&\geq C - C\left[\frac{1+M(\bar{S}-S^l)}{1+M(S^l-s)}\right]. \qquad (9.28)
\end{aligned}
$$

Since $M(t) \to \infty$ as $t \to \infty$, the expression in the square brackets in (9.28), which does not depend on S, goes to zero as $s \to -\infty$. Therefore, there is an $\bar{s} \leq S^l$, independent of S, such that $(1+M(\bar{S}-S^l))/(1+M(S^l-s)) < 1/2$ for all $s \leq \bar{s}$. This means that

$$
\mathcal{L}(s,S) > C/2 = \mathcal{L}(0,0) \text{ for all } s \leq \bar{s} \text{ and } S > s,
$$

and the proof is completed. □

REMARK 9.3 The proofs of Lemmas 9.3 and 9.4 can be easily extended to nondifferentiable renewal functions M by using Lebesgue-Stieltjes integrals. Therefore, the assertions of these two lemmas also hold for demand distributions which do not have densities.

THEOREM 9.1 *If the one-period demand has a density, then the function $\mathcal{L}(s,S)$, defined on $-\infty < S < \infty$, $s \leq S$, attains its minimum. Furthermore, if (s^*,S^*) is a minimum point of $\mathcal{L}(s,S)$, then $S^l \leq S^* \leq \bar{S}$ and $\bar{s} \leq s^* \leq S^*$.*

Proof. Because of Lemmas 9.3–9.5, the search for a minimum point can be restricted to the compact set $\{(s,S) : S^l \leq S \leq \bar{S},\ \bar{s} \leq s \leq S\}$. It immediately follows from the existence of a density of the one-period demand that $\mathcal{L}(s,S)$ is continuous, and therefore it attains its infimum over the compact set. □

It remains to show that the minimum is attained at an interior point and that there is a minimum point that satisfies (9.12).

LEMMA 9.6 *If $K > 0$, then there is a pair (s^*,S^*), with $s^* < S^*$, that minimizes $\mathcal{L}(s,S)$.*

Proof. We distinguish three cases.

- Case 1. If $m(x)$ is identically zero on $[0,\varepsilon]$ for some $\varepsilon > 0$, it immediately follows that $\tilde{\mathcal{L}}(\Delta,S)$ is constant for $\Delta \in [0,\varepsilon]$ and any fixed S. Therefore, if $(0,S^*)$ minimizes $\tilde{\mathcal{L}}(\Delta,S)$, so does (ε,S^*), which is an interior point.

- Case 2. Now let $m(0) > 0$. Any minimal point of $\hat{\mathcal{L}}(s, \Delta)$ satisfies (9.9). Let s^* be a solution of (9.9) for $\Delta = 0$. Then it follows from (9.10) that

$$\left.\frac{\partial \hat{\mathcal{L}}(s^*, \Delta)}{\partial \Delta}\right|_{\Delta=0} = -Km(0) < 0$$

  and therefore, $\Delta = 0$ cannot be a minimizer.

- Case 3. If $m(\cdot)$ does not satisfy either of the two cases above, it follows from the continuity of m as a sum of convolutions of continuous densities that there is an $\varepsilon > 0$ such that $m(x) > 0$ for all $x \in (0, \varepsilon]$. This property implies that $M(x) > 0$ for $x > 0$. For a fixed S, it holds that

$$\begin{aligned}
&\tilde{\mathcal{L}}(S, \Delta) - \tilde{\mathcal{L}}(S, 0) \\
&= \frac{K + L(S) + \int_0^{\Delta} L(S-x)m(x)dx}{1 + M(\Delta)} - (K + L(S)) \\
&= \frac{1}{1+M(\Delta)}(-M(\Delta)(K + L(S)) + L(S - \eta)M(\Delta)) \quad (9.29)
\end{aligned}$$

  for some $\eta \in [0, \Delta]$. Since $M(\Delta) > 0$ for $\Delta \in (0, \varepsilon]$ and L is continuous, it follows that for sufficiently small Δ

$$\tilde{\mathcal{L}}(S, \Delta) - \tilde{\mathcal{L}}(S, 0) < 0,$$

  and $\Delta = 0$ cannot be a minimizer. □

LEMMA 9.7 *There are s^* and S^*, with $s^* < S^*$, which minimize $\mathcal{L}(s, S)$ and, at the same time, satisfy* (9.12).

Proof. It follows from Lemma 9.6 that there is an interior minimizer $(s^\#, S^*)$ of $\mathcal{L}(s, S)$. If $m(S^* - s^\#) > 0$, it immediately follows from Iglehart's analysis that (9.12) is satisfied.

If $m(S^* - s^\#) = 0$, we define

$$\Delta_0 = \inf\{\Delta \geq 0 : m(x) \equiv 0 \text{ on } [\Delta, S^* - s^\#]\}$$

and

$$\Delta_1 = \sup\{\Delta \geq 0 : m(x) \equiv 0 \text{ on } [S^* - s^\#, \Delta]\}.$$

Obviously for $\Delta \in [\Delta_0, \Delta_1]$, the pair $(S^* - \Delta, S^*)$ minimizes $\mathcal{L}$. Let $\varepsilon > 0$. Then we have

$$\tilde{\mathcal{L}}(\Delta_1, S^*) - \tilde{\mathcal{L}}(\Delta_1 + \varepsilon, S^*) \leq 0.$$

Therefore, it follows that

$$\left(K + L(S^*) + \int_0^{\Delta_1} L(S^* - \xi)m(\xi)d\xi\right)(1 + M(\Delta_1 + \varepsilon)) \\ - \left(K + L(S^*) + \int_0^{\Delta_1+\varepsilon} L(S^* - \xi)m(\xi)d\xi\right)(1 + M(\Delta_1)) \leq 0$$

or

$$\left(K + L(S^*) + \int_0^{\Delta_1} L(S^* - \xi)m(\xi)d\xi\right)(M(\Delta_1 + \varepsilon) - M(\Delta_1)) \\ -(1 + M(\Delta_1))\int_{\Delta_1}^{\Delta_1+\varepsilon} L(S^* - \xi)m(\xi)d\xi \leq 0.$$

Applying the Mean Value Theorem A.1.9 and dividing by $(M(\Delta_1+\varepsilon) - M(\Delta_1))(1 + M(\Delta_1))$, which is strictly positive by the definition of Δ_1 and the monotonicity of $M(\cdot)$, we obtain

$$\frac{K + L(S^*) + \int_0^{\Delta_1} L(S^* - \xi)m(\xi)d\xi}{1 + M(\Delta_1)} \leq L(S^* - \eta)$$

for some $\eta \in [\Delta_1, \Delta_1 + \varepsilon]$. Since $L(y)$ is continuous, we find for $\varepsilon \to 0$,

$$\mathcal{L}(\Delta_1, S^*) - c\mu = \frac{K + L(S^*) + \int_0^{\Delta_1} L(S^* - \xi)m(\xi)d\xi}{1 + M(\Delta_1)} \leq L(S^* - \Delta_1). \tag{9.30}$$

If $\Delta_0 > 0$, we analogously find

$$\tilde{\mathcal{L}}(\Delta_0, S^*) \geq L(S^* - \Delta_0). \tag{9.31}$$

If $\Delta_0 = 0$, it is easy to see from (9.7) that

$$\tilde{\mathcal{L}}(0, S^*) - c\mu = K + L(S^*) > L(S^*). \tag{9.32}$$

Since $\tilde{\mathcal{L}}(\cdot, S^*)$ and $L(\cdot)$ are both continuous, it follows from (9.30)-(9.32) that there is a $\Delta^* > 0$ such that

$$\tilde{\mathcal{L}}(\Delta^*, S^*) - c\mu = L(S^* - \Delta^*),$$

i.e., the pair $(s^*, S^*) := (S^* - \Delta^*, S^*)$ is an interior minimizer that satisfies (9.12). □

Lemma 9.7 finally establishes the existence of an interior minimizer of the stationary cost function that satisfies equation (9.12) as required for Iglehart's analysis. With that analysis completed, the following result is easily established with the help of the same short additional argument that Derman suggested to Veinott and Wagner.

THEOREM 9.2 *The parameters s^* and S^*, obtained in Lemma 9.7, determine a stationary (s, S) strategy which is average optimal.*

REMARK 9.4 It follows from (9.9) that for any minimizer (s^*, S^*) of $\mathcal{L}(s, S)$, we have $s^* \leq \text{argmin } L(y)$. Therefore, for any two minimizers (s_1^*, S_1^*) and (s_2^*, S_2^*) of $\mathcal{L}(s, S)$ that satisfy (9.12), it holds that

$$L(s_1^*) = \mathcal{L}(s_1^*, S_1^*) - c\mu = k - c\mu = \mathcal{L}(s_2^*, S_2^*) - c\mu = L(s_2^*).$$

Since $k - c\mu > \min_S L(S)$, $L(x)$ is convex, and $\lim_{x\to\pm\infty} L(x) = \infty$, it follows that $s_1^* = s_2^*$.

9.8. Stationary Distribution Approach versus Dynamic Programming and Vanishing Discount Approach

In Chapters 5 and 6, we have established the average optimality of an (s, S) strategy in the more general setting of Markovian demand. We use dynamic programming and a vanishing discount approach to obtain the average cost optimality equation and show that it has a K-convex solution, which provides an (s, S) strategy that minimizes (9.2). Furthermore, we prove a verification theorem stating that any stable policy (defined later in the section; see (9.35)) satisfying the average cost optimality equation is average optimal.

More specifically, we prove that there is a policy Y^* that minimizes the average cost defined by

$$J(x|Y) = \limsup_{n\to\infty} \frac{1}{n} f_n(x|Y)$$

over all history-dependent policies $Y \in \mathcal{Y}$. Furthermore, this policy Y^* can be represented as an (s, S) policy. In addition, we show that this policy also minimizes the criterion

$$a(x|Y) = \liminf_{n\to\infty} \frac{1}{n} f_n(x|Y)$$

over the class of all stable policies.[4]

Moreover, the completion of Iglehart's analysis in the previous section allows us to drop the stability restriction on the class of admissible strategies and to obtain the stronger result that an optimal (s, S) strategy also minimizes $a(x|Y)$ over *all* history-dependent policies $Y \in \mathcal{Y}$. Theorem 9.3 that follows connects the two approaches.

For the average cost problem under consideration, the average cost optimality equation derived in Chapter 6 is given by

$$\psi(x) = \min_{y \geq x}[\hat{c}(y - x) + L(y) - \lambda + \int_0^\infty \psi(y - \xi)\varphi(\xi)d\xi]. \tag{9.33}$$

A pair (λ^*, ψ^*) such that

$$\psi^*(x) = \min_{y \geq x}[\hat{c}(y - x) + L(y) - \lambda^* + \int_0^\infty \psi^*(y - \xi)\varphi(\xi)d\xi]$$

is called a *solution* of (9.33). Note that for $\lambda = k$, (9.33) reduces to equation (9.19), specified by Iglehart.

THEOREM 9.3 *Let (λ^*, ψ^*) be a solution of the average cost optimality equation* (9.33). *Let ψ^* be continuous and let the minimizer on the RHS of* (9.33) *be given by*

$$y(x) = \begin{cases} S^* & \text{if } x \leq s^*, \\ x & \text{if } x > s^*, \end{cases}$$

for $-\infty < s^* \leq S^* < \infty$. Then,

(a) *the pair (s^*, S^*) minimizes $\mathcal{L}(s, S)$,*

(b) (s^*, S^*) *satisfies* (9.12), *and*

(c) *for all history-dependent policies $Y \in \mathcal{Y}$, it holds that*

$$\lambda^* = k = \lim_{n\to\infty} \frac{1}{n} f_n(x|s^*, S^*) \leq \liminf_{n\to\infty} \frac{1}{n} f_n(x|Y).$$

---

[4]Bounds on the action space imposed by Tijms (1972) imply that the admissible policies he considers are stable. In this case, either of the criteria — lim inf or lim sup — can be used in the MDP context.

Proof. To prove (a) we assume to the contrary that the pair (s^*, S^*) does not minimize $\mathcal{L}(s, S)$. Then, there is another strategy (s, S) with $\mathcal{L}(s^*, S^*) > \mathcal{L}(s, S)$. Therefore, in view of Remark 9.1, we obtain

$$\limsup_{n\to\infty} \frac{1}{n} f_n(x|s^*, S^*) = \mathcal{L}(s^*, S^*) > \mathcal{L}(s, S) = \limsup_{n\to\infty} \frac{1}{n} f_n(x|s, S). \tag{9.34}$$

Equation (9.34) contradicts the optimality of (s^*, S^*) proved in Chapter 6 for the average cost objective function (9.2). Therefore, (s^*, S^*) minimizes $\mathcal{L}(s, S)$.

It is shown in Chapter 6 that the λ^* from the solution of the average cost optimality equation is equal to the minimum of the average cost defined in (9.2), and is therefore equal to $k = \mathcal{L}(s^*, S^*)$. Knowing this, part (c) of Theorem 9.3 immediately follows from (a) and Theorem 9.2.

For the proof of (b), we note that ψ^* can by expressed as in (9.20). As shown in Iglehart, ψ^* is continuous if and only if (9.12) holds. This proves part (b) of Theorem 9.3. □

THEOREM 9.4 *If* $\lim_{x\to-\infty}[cx+L(x)] = \infty$, *then there is a unique (up to a constant) continuous bounded-from-below* K*-convex function* ψ^* *such that* (λ^*, ψ^*) *is a solution of the average cost optimality equation* (9.33). *Furthermore,* λ^* *is equal to the minimal average cost with respect to either* (9.1) *or* (9.2).

Proof. Because of the K-convexity of the solution, the minimizer on the RHS of (9.19) is given by

$$y(x) = \begin{cases} S^* & \text{if} \quad x \le s^*, \\ x & \text{if} \quad x > s^*, \end{cases}$$

for some not necessarily finite $s^* \le S^*$. Since $\lim_{x\to-\infty}(cx + L(x)) = \infty$ and ψ^* is bounded from below, it follows that

$$\lim_{x\to-\infty} \left[cx + L(x) - k + \int_0^\infty \psi^*(x - \xi)\varphi(\xi)d\xi\right] = \infty,$$

and therefore, $s^* > -\infty$. Since $\lim_{x\to\infty} L(x) = \infty$ and ψ^* is bounded from below, it is obvious that

$$\lim_{y\to\infty} \left[cy + L(y) - k + \int_0^\infty \psi^*(y - \xi)\varphi(\xi)d\xi\right] = \infty,$$

and therefore, $S^* < \infty$.

It is easy to show that the (s, S) strategy with finite parameters s^* and S^* is *stable with respect to* ψ^*, i.e.,

$$\lim_{n\to\infty} \frac{\psi^*(x_n)}{n} = 0. \tag{9.35}$$

Given this fact, it is proved in Chapter 6 that (s^*, S^*) is an optimal strategy with respect to the average cost defined in (9.2) with minimal average cost λ^*. It follows from Theorem 9.3 that $\lambda^* = k$.

By Theorem 9.3, (s^*, S^*) minimizes $\mathcal{L}(s, S)$ and satisfies (9.12). It follows from Remark 9.4 that s^* is unique. The K-convexity and the parameter s^* uniquely determine the solution (9.20) of (9.19) (up to a constant). Therefore, (λ^*, ψ^*) is the unique solution of (9.33) with the desired properties. □

REMARK 9.5 In the case of a constant unit shortage cost p, the condition $\lim_{x\to-\infty}(cx + L(x)) = \infty$ is equivalent to the requirement $p > c$. This condition is only introduced to simplify the proof and can be dropped altogether.

9.9. Concluding Remarks and Notes

In this chapter, based on Beyer and Sethi (1999), we have reviewed the classical papers of Iglehart (1963b) and Veinott and Wagner (1965), treating single-product average cost inventory problems. We have pointed out some conditions that are assumed implicitly but not proved in these papers, and we have proved them rigorously. In particular, we have provided bounds for any pair (s^*, S^*) minimizing the stationary one-period cost $\mathcal{L}(s, S)$, using only the properties of the stationary cost function.

The main purpose of our analysis in this chapter has been to complete the stationary distribution analyses of Iglehart and Veinott and Wagner and relate it to the vanishing discount approach. Therefore, we have stayed with the relatively restrictive assumptions on the demand distribution made by Iglehart. However, these assumptions are not necessary for the results obtained in this chapter. Indeed, it can be shown that even for general demand distributions consisting of discrete and continuous components, the corresponding stationary cost function attains its minimum provided the expected values of all the quantities required in the analysis exist. Furthermore, there is a pair minimizing $\mathcal{L}(s, S)$ that satisfies (9.12) and that can in turn be proved to be average optimal.

We have also introduced the connection between the stationary distribution approach and the dynamic programming approach to the problem. While dynamic programming is applicable even in problems defying a stationary analysis but with objective function (9.2), the stationary analysis, when possible, can prove optimality with respect to the more conservative objective function (9.1). Finally, by combining both approaches, we have shown that an (s, S) policy – optimal with respect to either of the objective functions (9.1) or (9.2) – is also optimal with respect to the other.

Finally, we mention that Presman and Sethi (2006) have developed a unified stationary distribution approach that deals with both the discounted and long-run average cost problems in continuous time. Bensoussan *et al.* (2005b), on the other hand, use the dynamic programming approach in the continuous-time case, which use the theory of quasivariational inequalities. They only deal with the discounted problem, however.

V

CONCLUSIONS AND OPEN RESEARCH PROBLEMS

Chapter 10

CONCLUSIONS AND OPEN RESEARCH PROBLEMS

The Markovian demand approach provides a realistic way of modeling real-world demand scenarios. It allows us to relax the common assumption of demands being independent over time in the inventory literature. By associating the demand process with an underlying Markov chain, we are able to capture the effect of environmental factors that influence the demand process. Although the modeling capability is significantly enhanced by the incorporation of Markovian demands in inventory models, the simplicity of the optimal policies normally exhibited in the classical inventory problems is still preserved. Specifically, we show that the (s, S)-type policies shown to be optimal for a large class of inventory models with independent demands continue to be optimal for Markovian demand models, with one difference. That is, with Markovian demands, the (s, S) values depend on the state of the Markov process.

In this book, we provide a comprehensive treatment for various inventory models with Markovian demands. We treat the discounted cost models as well as the long-run average cost models. Both the backlog and the lost sales cases are treated. Moreover, the assumptions on the demand and various costs are generalized. The results are derived in a mathematically rigorous manner. For each model presented, we use dynamic programming to prove the existence of an optimal feedback policy before analyzing its specific structure. For the analysis of long-run average cost models, we use the vanishing discount approach.

In Chapter 9, we provide a complete analysis of the classical inventory model with a fixed cost by using the stationary distribution approach as a way of contrasting with the vanishing discount approach throughout the book. For this analysis, we use i.i.d. demands.

D. Beyer et al., *Markovian Demand Inventory Models*, International Series in Operations Research & Management Science 108,
DOI 10.1007/978-0-387-71604-6_10, © Springer Science+Business Media, LLC 2010

Some of the results derived for the Markovian demand models require nontrivial extensions of the existing results in the literature. An example is the generalization of the concept of K-convexity and its related properties. The generalization allows us to also relax some of the restrictive assumptions used in the literature.

We have introduced Markov decision processes (MDP) as a way of modeling demand when demand is endogenous. This can happen when the advertising or promotion is used to increase the demand as in Chapter 8. It would be of great interest to include pricing decisions in such models. We believe our attempt is just a starting point toward using the rich literature on the MDP theory for analysis of realistic inventory models.

Some of the open problems have been discussed in the concluding remarks and end notes of the relevant chapters. In addition, there are many possible, as well as interesting, extensions of the works presented in this book. Song and Zipkin (1993) have analyzed an inventory model that corresponds to a continuous-time version of the model presented in Chapter 2. It would be of interest to study continuous-time models corresponding to other models analyzed here. In this vein, Presman and Sethi (2006) and Bensoussan *et al.* (2005b) have proved the optimality of the (s, S) policy in inventory models that incorporate constant and/or diffusion demands, along with the standard compound Poisson demand. Extending these works to allow for a Markovian demand structure remains an open problem.

A second set of extensions concerns the inclusion of pricing and other marketing activities in the MDP framework, along the lines pursued in Chapter 8.

In the literature, there have been some attempts to develop procedures to compute the (s, S) policy parameters defining the optimal policy. Earlier computational methods based on the policy iteration algorithm for dynamic programming problems proved to be prohibitively expensive because of the computational complexity involved in the two-dimensional search for the optimal values of s and S. For example, the methods of Veinott and Wagner (1965) essentially apply a full enumeration on the two-dimensional grid of s and Δ ($\Delta = S - s$), facilitated by the bounds identified for the optimal s and S. Zheng and Federgruen (1991) provide the first efficient algorithm for computing the optimal (s, S) policy by reducing the dimensionality of the modified policy iteration to one dimension in each iteration. To compute state-dependent (s, S) policy parameters that arise in this book is a much harder problem, as well as an interesting research area.

It continues to be a challenging open research problem to find optimal policy structures in inventory problems with multiple products that are interrelated by way of joint setup costs, warehousing constraints, joint demand distributions, etc. Some attempts involving multiple products with i.i.d. demands, but connected by a warehouse constraint, include Beyer and Sethi (1997) and Beyer *et al.* (2001), and references therein. These could be extended to allow for Markovian demands. Recently, Gallego and Sethi (2005) have generalized the concept of K-convexity to n-dimensional spaces. An obvious next step is to see how these concepts could be used for analysis of inventory problems with related multiple products, possibly with Markovian demands.

Recently, with the realization that demands and inventories may not be fully observed even after they are realized; (see, e.g., Raman *et al.* (2001)), there has been a great deal of interest in studying problems with incomplete inventory information. Bensoussan et al. (2007c,2008b) have formulated models where signals such as presence or absence of inventories are observed instead of full inventory levels. For these models with partial observations, they derive appropriate dynamic programming equations in infinite dimensional spaces and prove the existence of optimal feedback policies in terms of the conditional distribution of the inventory level given the past signal observations. Sometimes, full observation comes after a delay, in which case Bensoussan *et al.* (2007b) have shown a modified (s, S) policy to be optimal in the case when the information delays are modeled as a Markov process. A generalization of these works to allow for environmental states remains open.

Ding *et al.* (2002), Lu *et al.* (2005), Bensoussan et al. (2009a,2009b) have considered censored demand models when the demands are i.i.d. with an unknown parameter, whereas Bensoussan et al. (2007a,2008a) allow for demands to be Markov processes. While optimal policies in these models are much more complicated, they can still be extended to allow for environmental states. Furthermore, these environmental states may also be only partially observed.

VI

APPENDICES

Appendix A

ANALYSIS

The definitions and results presented here can be found in standard texts on analysis such as Rudin (1976), Friedman (1970), Krantz (2004), Ross (1980), and Yosida (1980). See Rockafellar (1970) for concepts and results related to Lipschitz continuity. Internet sources that deal with these topics are websites such as mathworld.wolfram.com, answers.com, en.wikipedia.org, and planetmath.org.

A.1. Continuous Functions on Metric Spaces

We consider only real-valued functions defined on a metric space X endowed with a metric ρ. Thus, $\rho(x, y)$ denotes the *distance* between $x, y \in X$. Throughout the book, the cases of X that we are interested in are $\mathbb{R}$, $\mathbb{R}^n$ and $\mathbb{I} \times \mathbb{R}^n$, where $\mathbb{I}$ is a finite set.

Definition A.1.1 (Continuity) A function f defined on a subset A of a metric space X is said to be *continuous* on A if for any $\varepsilon > 0$ and $x \in A$, there is a $\delta > 0$ such that $|f(y) - f(x)| < \varepsilon$ whenever $y \in A$ with $\rho(y, x) < \delta$.

Definition A.1.2 (Lower Semicontinuity) A function f defined on a metric space X is said to be *lower semicontinuous* (l.s.c.) if for any $x \in X$,

$$\liminf_{y \to x} f(y) \geq f(x).$$

Furthermore, f is *upper semicontinuous* (u.s.c.) if the function $-f$ is l.s.c.

Example. Any increasing function on real line is l.s.c. if it is continuous from the left, and it is u.s.c. if it is continuous from the right. In particular, a cumulative distribution function Φ of any random variable ξ is always u.s.c.

Theorem A.1.1 *A function f defined on $[a, b] \subset \mathbb{R} \cup \{-\infty, \infty\}$ is l.s.c. if and only if the set $\{x \in [a, b] | f(x) \leq M\}$ is closed for each M.*

D. Beyer et al., *Markovian Demand Inventory Models*, International Series in Operations Research & Management Science 108,
DOI 10.1007/978-0-387-71604-6,

DEFINITION A.1.3 (Uniform Continuity) A function f, defined on a subset A of a metric space X, is said to be *uniformly continuous* on A if for any $\varepsilon > 0$ there is a $\delta > 0$ such that $|f(y) - f(x)| < \varepsilon$ whenever y and x belong to A and $\rho(y, x) < \delta$.

THEOREM A.1.2 *If f is a continuous function on a compact subset A of a metric space X, then f is uniformly continuous on A.*

DEFINITION A.1.4 (Lipschitz Continuity) A function f defined on a subset A of a metric space X is *Lipschitz continuous* if there is a constant $C \geq 0$ such that $|f(y) - f(x)| < C\rho(y, x)$ for all x and y in A. Furthermore, the smallest such C is called the *Lipschitz constant* of f.

DEFINITION A.1.5 (Local Lipschitz Continuity) A function f defined on $\mathbb{R}$ is *locally Lipschitz continuous* if for any $b > 0$ there is a constant $C \geq 0$ such that $|f(y) - f(x)| < C|y - x|$ for all $x, y \in [-b, b]$.

THEOREM A.1.3 *Every locally Lipschitz continuous function on $\mathbb{R}$ is uniformly continuous on any finite interval of $\mathbb{R}$.*

THEOREM A.1.4 (Kirszbraun Theorem) *If $A \subset \mathbb{R}^n$ and $f : A \to \mathbb{R}^m$ is Lipschitz continuous, then there is a Lipschitz continuous function $F : \mathbb{R}^n \to \mathbb{R}^m$ that extends f, and has the same Lipschitz constant as f. Moreover, if f is locally Lipschitz continuous, then the extension F is also locally Lipschitz continuous.*

THEOREM A.1.5 (Existence Theorem) *A l.s.c. function from $\mathbb{R}^n \to \mathbb{R}$, which is bounded from below, attains its minimum on a compact subset of $\mathbb{R}^n$.*

THEOREM A.1.6 *Let $c(x, u)$, a function from $\mathbb{R}^n \times \mathbb{R}^+$ into $\mathbb{R}$, be l.s.c. Then, the function $\nu : x \to \nu(x) = \inf_{u \geq 0} c(x, u)$ is l.s.c.*

THEOREM A.1.7 (Selection Theorem) *Let $c(x, u)$, a function from $\mathbb{R}^n \times \mathbb{R}^m$ into $\mathbb{R}$, be l.s.c. and bounded from below. Consider*

$$\inf_{u \in A} c(x, u),$$

where A is a compact subset in $\mathbb{R}^m$. Then, the minimum is attained on a bounded subset of points in A, which depends on x. Furthermore, there exists a Borel map $u^*(x) : \mathbb{R}^n \to \mathbb{R}^m$ such that

$$c(x, u^*(x)) = \inf_{u \in A} c(x, u).$$

The following theorem shows that the property of l.s.c. is also preserved under the monotone limit procedure.

THEOREM A.1.8 *Let $f_1(x) \leq f_2(x) \leq \ldots$, be a sequence of l.s.c. functions on $\mathbb{R}^n$, and suppose that for each $x \in \mathbb{R}^n$,*

$$\lim_{n \to \infty} f_n(x) = f(x).$$

Then, the function $f(x)$ is l.s.c. In particular, if $\{g_n(x)\}$ is a sequence of nonnegative l.s.c. functions such that

$$\sum_{n=1}^{\infty} g_n(x) = g(x),$$

then the function $g(x)$ is l.s.c.

THEOREM A.1.9 (Mean Value Theorem) *If $f : [a, b] \to \mathbb{R}$ is continuous on $[a, b]$ and differentiable on (a, b), then there is a point $x \in (a, b)$ at which*

$$f(b) - f(a) = (b - a) f'(x).$$

A.2. Convergence of a Sequence of Functions

In this section, we consider only real-valued functions defined on $\mathbb{R}$.

DEFINITION A.2.1 (Pointwise Convergence) A sequence of functions $\{f_n\}$ is said to be *pointwise convergent* to a function f on a set A if for any $x \in A$ and any $\varepsilon > 0$, there exists an integer $N_{\varepsilon,x}$ such that $|f_n(x) - f(x)| < \varepsilon$ for $n \geq N_{\varepsilon,x}$.

DEFINITION A.2.2 (Uniform Convergence) A sequence of functions $\{f_n\}$ is said to be *uniformly convergent* to a function f on a set A if for each $\varepsilon > 0$, there corresponds an integer N_ε such that $|f_n(x) - f(x)| < \varepsilon$ for each $x \in A$ and $n \geq N_\varepsilon$.

THEOREM A.2.1 (Dini's Theorem) *If an increasing sequence of continuous functions $\{f_n\}$ is pointwise convergent to a continuous function f on a compact set A, then it is uniformly convergent to f on A.*

THEOREM A.2.2 *Let $\{f_n\}$ be a sequence of continuous functions on $[a, b]$ that converges uniformly to a function f on $[a, b]$. Let $\{x_n\}$ be a sequence in $[a, b]$ converging to x. Then,*

$$\lim_{n \to \infty} f_n(x_n) = f(x).$$

DEFINITION A.2.3 (Local Uniform Convergence) A sequence of functions $\{f_n\}$ defined on $\mathbb{R}$ is said to be *locally uniformly convergent* to a function f if for any $b > 0$ and $\varepsilon > 0$, there corresponds an integer $N_{\varepsilon,b}$ such that $|f_n(x) - f(x)| < \varepsilon$ for each $x \in [-b, b]$ and $n \geq N_{\varepsilon,b}$.

THEOREM A.2.3 (Uniform Convergence Theorem) *If $\{f_n\}$ is a sequence of continuous functions on a set A that converges uniformly to a function f on the set A, then f is continuous on the set A.*

The theorem below allows us to exchange integrals and limit processes.

THEOREM A.2.4 *If $\{f_n\}$ is a sequence of integrable functions on a bounded interval $[a, b]$ that converge uniformly to a limit f on $[a, b]$, then f is integrable and*

$$\int_a^b f = \lim_{n \to \infty} \int_a^b f_n.$$

THEOREM A.2.5 *Let $\{f_n\}$ be a sequence of functions differentiable on an interval $[a, b]$, converging pointwise to a function f. Suppose that the derivative f_n' is continuous for all n, and that the differentiated sequence converges uniformly to a function g. Then, the sequence $\{f_n\}$ converges uniformly to f on $[a, b]$ and $f' = g$.*

A.3. The Arzelà-Ascoli Theorems

DEFINITION A.3.1 (Uniform Boundedness) A family $\{f_\beta\}$ of continuous functions on a subset $A \subset \mathbb{R}$ is said to be *uniformly bounded* if there is a constant $C > 0$ such that $|f_\beta(x)| \leq C$ for all β and for all $x \in A$.

DEFINITION A.3.2 (Equicontinuity) A family $\{f_\beta\}$ of functions defined on a subset $A \subset \mathbb{R}$ is said to be *equicontinuous* if for any given $\varepsilon > 0$ and $x \in A$, there exists a $\delta > 0$ such that

$$|f_\beta(y) - f_\beta(x)| < \varepsilon$$

for all y in A, for which $|y - x| < \delta$ and for all β.

DEFINITION A.3.3 (Uniform Equicontinuity) A family $\{f_\beta\}$ of functions defined on a subset $A \subset \mathbb{R}$ is said to be *uniformly equicontinuous* if for any given $\varepsilon > 0$, there exists a $\delta > 0$ such that

$$|f_\beta(y) - f_\beta(x)| < \varepsilon$$

for each x, y in A, for which $|y - x| < \delta$ and for all β.

For comparison, the statement that all functions in $\{f_\beta\}$ are *continuous* on $A \subset \mathbb{R}$ means that for every $\varepsilon > 0$, every β, and every $x \in A$, there exists a $\delta > 0$ such that for $y \in A$ with $|y - x| < \delta$, we have $|f_\beta(y) - f_\beta(x)| < \varepsilon$. So, for continuity, δ may depend on ε, x, and β. For uniform continuity, δ may depend on ε and β, but not on x; for equicontinuity, δ may depend on ε and x, but not on β; and for uniform equicontinuity, δ may depend on ε, but not on β and x.

DEFINITION A.3.4 (Equi-Lipschitz Continuity) A family $\{f_\beta\}$ of functions defined on a subset $A \subset \mathbb{R}$ is *equi-Lipschitzian* on A if there exists a $C \geq 0$ such that

$$|f_\beta(y) - f_\beta(x)| < C|y - x|$$

for all x, y in A and all β.

THEOREM A.3.1 *If a family $\{f_\beta\}$ of functions is equi-Lipschitzian on an interval $[a, b] \in \mathbb{R}$, then $\{f_\beta\}$ is uniformly equicontinuous on $[a, b]$.*

DEFINITION A.3.5 (Local Equi-Lipschitz Continuity) A family $\{f_\beta\}$ of functions defined on $\mathbb{R}$ is *locally equi-Lipschitzian* if for any $b > 0$, there exists a $C \geq 0$ such that

$$|f_\beta(y) - f_\beta(x)| < C|y - x|$$

for all x, y in $[-b, b]$ and all β.

THEOREM A.3.2 *Let $\{f_\beta\}$ be a locally equi-Lipschitzian family of functions defined on $\mathbb{R}$. Then $\{f_\beta\}$ is uniformly equicontinuous on any compact subset of $\mathbb{R}$.*

Properties of boundedness, continuity, Lipschitz continuity, and local Lipschitz continuity are preserved under pointwise convergence if they are satisfied uniformly with the same constants for the converging sequence. Namely, the following result holds.

THEOREM A.3.3 *If $\{f_n\}$ is a sequence of equicontinuous, locally equi-Lipschitzian, equi-Lipschitzian, or uniformly bounded functions on a set A that converge pointwise to a function f on the set A, then f is continuous, Lipschitzian, locally Lipschitzian, or bounded, respectively on A.*

THEOREM A.3.4 (Arzelà-Ascoli Theorem) *Let $\{f_\beta\}$ be a family of functions, uniformly bounded and uniformly equicontinuous on a compact subset $A \subset \mathbb{R}$. Then, any sequence $\{f_n\}$ of functions of $\{f_\beta\}$ has a subsequence that is uniformly convergent on A to a continuous function f on A.*

In this book, we use the following version of the Arzelà-Ascoli Theorem, which can easily be derived from Theorems A.3.3, A.3.2, and A.3.4.

THEOREM A.3.5 (Arzelà-Ascoli Theorem) *Let $\{f_\beta\}$ be a family of functions, uniformly bounded and locally equi-Lipschitz continuous. Then, any sequence $\{f_n\}$ of functions of $\{f_\beta\}$ has a subsequence that is locally uniformly convergent to a locally Lipschitz continuous function f.*

The sketch of the proof of this theorem is the following. Take a sequence of intervals $A_N = [-N, N], N = 1, 2 \ldots$. Let $\{f_{1,n_k}\}, k = 1, 2 \ldots$ be a subsequence of $\{f_n\}$, which is uniformly convergent on A_1. Choose $\{m, n_k\}, k = 1, 2 \ldots$ a subsequence of $\{(m-1), n_k\}$ in such a way that $\{f_{m,n_k}\}$ converges uniformly on A_m. The sequence of functions $\{f_{N,n_N}\}, N = 1, 2 \ldots$ converges uniformly on any A_k, i.e., it is locally uniformly convergent. The rest follows from Theorem A.3.3.

A.4. Linear Operators

DEFINITION A.4.1 (Linear Operator) Let X and Y be linear vector spaces. A function T from a subset $A \subset X$ into Y is called an *operator.* The set A where T is defined is called the *domain* of T. Furthermore, T is linear if A is a linear subset of X and if

$$T(\lambda_1 x_1 + \lambda_2 x_2) = \lambda_1 T x_1 + \lambda_2 T x_2$$

for all x_1, x_2 in A and λ_1, λ_2 scalars.

DEFINITION A.4.2 (Continuous Linear Operator) Let X and Y be normed linear spaces. A *continuous linear operator* T from X into Y is a linear operator that is also continuous – that is, if $x_n \to x$, then $Tx_n \to Tx$.

Let X and Y be normed linear spaces, and let T be a linear operator from X into Y. If there is a constant K such that, for all $x \in X$,

$$||Tx|| \leq K||x||,$$

then we say that T is *bounded* and we call T a bounded linear operator. Note that the norms on the different sides of the last inequality are taken in different spaces. Furthermore, the *norm* of the linear operator T, denoted as $||T||$, is defined as

$$||T|| = \sup_{x \neq 0} \frac{||Tx||}{||x||} = \sup_{||x||=1} ||Tx||.$$

THEOREM A.4.1 *A linear operator T from a normed linear space X into a normed linear space Y is continuous if and only if it is bounded.*

For the operator F_{k+1} defined in (3.8), we have the following result.

LEMMA A.4.1 *F_{k+1} is a continuous linear operator from $\mathbb{B}^\gamma$ into $\mathbb{B}^\gamma$.*

Proof. It immediately follows from the definition of F_{k+1} that it is a linear operator from $\mathbb{B}^\gamma$ into $\mathbb{B}^\gamma$. By definition of $\| \ b \ \|_\gamma$ we have

$|b(i,x)| \leq \| b \|_\gamma (1+|x|^\gamma)$, and we can conclude that

$$\begin{aligned}
\| F_{k+1} \| &= \sup_{b\neq 0} \frac{\| F_{k+1}b \|_\gamma}{\| b \|_\gamma} \\
&= \sup_{b\neq 0} \frac{\| \sum_{j=1}^{L} p_{ij} \int_0^\infty b(j,y-\xi) d\Phi_{i,k}(\xi) \|_\gamma}{\| b \|_\gamma} \\
&\leq \sup_{b\neq 0} \frac{\| \sum_{j=1}^{L} p_{ij} \int_0^\infty \| b \|_\gamma (1+|y-\xi|^\gamma) d\Phi_{i,k}(\xi) \|_\gamma}{\| b \|_\gamma} \\
&= \| \sum_{j=1}^{L} p_{ij} \int_0^\infty (1+|y-\xi|^\gamma) d\Phi_{i,k}(\xi) \|_\gamma \\
&= \max_i \sup_y \frac{| \sum_{j=1}^{L} p_{ij} \int_0^\infty (1+|y-\xi|^\gamma) d\Phi_{i,k}(\xi)|}{1+|y|^\gamma} \\
&\leq 1 + \max_i \sup_y \frac{\sum_{j=1}^{L} p_{ij} \int_0^\infty (|y|+|\xi|)^\gamma d\Phi_{i,k}(\xi)}{1+|y|^\gamma}.
\end{aligned}$$

Using the inequality $(a+b)^\gamma \leq 2^\gamma(|a|^\gamma + |b|^\gamma)$, we obtain

$$\begin{aligned}
\| F_{k+1} \| &\leq 1 + 2^\gamma \max_i \sup_y \frac{\sum_{j=1}^{L} p_{ij} (|y|^\gamma + \int_0^\infty |\xi|^\gamma d\Phi_{i,k}(\xi))}{1+|y|^\gamma} \\
&\leq 1 + 2^\gamma (1 + \max_i \int_0^\infty |\xi|^\gamma d\Phi_{i,k}(\xi)) \\
&= 1 + 2^\gamma \Big(1 + \max_i \mathsf{E}\Big[|\xi_k|^\gamma \Big| i_k = i\Big]\Big),
\end{aligned}$$

which is finite on account of (3.1). Therefore, the operator F_{k+1} is bounded and therefore continuous. □

A.5. Miscellany

Definition A.5.1 A sequence of $\{a_n\}_{n=0}^\infty$ is said to be *Cesaro-summable* to the limit s if

$$\lim_{N\to\infty} S_N = s,$$

where

$$S_N = \frac{1}{N+1} \sum_{n=0}^{N} a_n.$$

A sequence of $\{a_n\}_{n=0}^\infty$ is said to be *Abel-summable* to a if

$$\lim_{\beta\to 1^-} (1-\beta) \sum_{n=0}^{\infty} \beta^n a_n = a.$$

THEOREM A.5.1 *If a sequence of real numbers* $\{a_n\}_{n=0}^{\infty}$ *is Cesaro-summable to* s, *then it is also Abel-summable to* s.

THEOREM A.5.2 (The Tauberian Theorem) *Consider an arbitrary sequence* $\{a_n\}_{n=0}^{\infty}$ *for which the function*

$$f(\beta) := (1-\beta)\sum_{n=0}^{\infty} \beta^n a_n$$

is well defined for all $|\beta| < 1$, *that is, the* RHS *in the above expression converges absolutely. Then, we have*

$$\liminf_{N\to\infty} S_N \leq \liminf_{\beta\to 1^-} f(\beta) \leq \limsup_{\beta\to 1^-} f(\beta) \leq \limsup_{N\to\infty} S_N. \tag{A.1}$$

See Sznajder and Filar (1992), Theorem 2.2.

Appendix B

PROBABILITY

In Sections B.1 and B.2, we define concepts and related results used in this book. In Section B.3, we state the Elementary Renewal Theorem. See standard references such as Chow and Teicher (1978), Chung (1974), and Kallenberg (2001), and websites such as en.wikipedia.org, mathworld.wolfram.com, and planetmath.org. Section B.4 is concerned with concepts of stochastic dominance, for which Shaked and Shanthikumar (1984) is a good reference.

Let $(\Omega, \mathcal{F}, \mathsf{P})$ denote a probability space. Let $X = X(\omega), \omega \in \Omega$, denote a random variable defined on the probability space. Two random variables X and Y are said to be equivalent (or to coincide) if $X = Y$ a.s. Let $\mathcal{B}$ be a sub σ-algebra of $\mathcal{F}$.

B.1. Integrability

DEFINITION B.1.1 (Integrability) A random variable X is said to be *integrable* if

$$\mathsf{E}|X| = \int |X| d\mathsf{P} < \infty.$$

DEFINITION B.1.2 (Uniform Integrability) A sequence X_n of random variables is called *uniformly integrable* if

$$\sup_{n \geq 1} \mathsf{E}|X_n| d\mathsf{P} < \infty;$$

and in addition, for any $\varepsilon > 0$, there exists a $\delta > 0$ such that

$$\sup_{n \geq 1} \int_A |X_n| d\mathsf{P} < \varepsilon,$$

whenever $\mathsf{P}(A) < \delta$. Or, equivalently, if

$$\lim_{\beta \to \infty} \sup_{n \geq 1} \int_{\{|X_n| > \beta\}} |X_n| d\mathsf{P} = 0.$$

LEMMA B.1.1 (Fatou's Lemma) *Let $\{X_n\}$ be a sequence of nonnegative integrable random variables. Then,*

$$\mathsf{E} \liminf_{n\to\infty} X_n \leq \liminf_{n\to\infty} \mathsf{E} X_n.$$

B.2. Conditional Expectation

DEFINITION B.2.1 (Conditional Expectation) The *conditional expectation* of the random variable X with respect to $\mathcal{B}$ is the unique (up to equivalence) $\mathcal{B}$-measurable random variable, denoted as $\mathsf{E}(X|\mathcal{B})$, such that

$$\mathsf{E}[\mathsf{E}(X|\mathcal{B})Y] = \mathsf{E}XY$$

for any bounded $\mathcal{B}$-measurable random variable Y. Alternatively, we define $\mathsf{E}(X|\mathcal{B})$ to be any $\mathcal{B}$-measurable random variable such that

$$\int_B \mathsf{E}(X|\mathcal{B})d\mathsf{P} = \int_B X d\mathsf{P}, \quad \forall B \in \mathcal{B}.$$

Furthermore, the conditional expectation of X with respect to a random variable Z, denoted as $\mathsf{E}(X|Z)$, is equal to its conditional expectation with respect to the σ-algebra generated by Z.

B.2.1 Properties of Conditional Expectation

(i) $X \geq 0$ implies $\mathsf{E}(X|\mathcal{B}) \geq 0$ a.s.;

(ii) $\mathsf{E}|\mathsf{E}(X|\mathcal{B})| \leq \mathsf{E}|X|$;

(iii) $0 \leq X_n \uparrow X$ implies $\mathsf{E}(X_n|\mathcal{B}) \uparrow \mathsf{E}(X|\mathcal{B})$ a.s.;

(iv) $\mathsf{E}(XY|\mathcal{B}) = Y\mathsf{E}(X|\mathcal{B})$ a.s. when Y is $\mathcal{B}$-measurable;

(v) $\mathsf{E}(Y\mathsf{E}(X|\mathcal{B})) = \mathsf{E}(X\mathsf{E}(Y|\mathcal{B})) = \mathsf{E}(\mathsf{E}(X|\mathcal{B})\mathsf{E}(Y|\mathcal{B}))$;

(vi) $\mathsf{E}(\mathsf{E}(X|\mathcal{B})|\mathcal{C})) = \mathsf{E}(X|\mathcal{C})$ a.s. for all $\mathcal{C} \subset \mathcal{B}$;

(vii) Let X and Y be two random variables and $g : \mathbb{R}^2 \to \mathbb{R}$ be a measurable function such that $\mathsf{E}|g(X,Y)| < \infty$. Let Y be $\mathcal{B}$-measurable. Let $h(\omega, y)$ be a measurable function on $\Omega \times \mathbb{R}$ such that for each $y \in \mathbb{R}$,

$$h(\omega, y) = \mathsf{E}(g(X, y)|\mathcal{B}).$$

Then, we have

$$\mathsf{E}(g(X, Y)|\mathcal{B}) = h(\omega, Y).$$

B.3. Renewal Theorem

Let $\{X_n\}$ be a sequence of nonnegative i.i.d. random variables, with X_n denoting time between the $(n-1)$st and n^{th} event and $\mu = \mathsf{E}X_n$. Let

$$S_n = \sum_{i=1}^{n} X_i, \quad n \geq 1,$$

and $N(t) = \max\{n | S_n \leq t\}$, the number of events by time t. It is easy to verify that $N(t)+1$ is a stopping time of the sequence $\{X_n\}$, that is, the event $N(t)+1 = k$ belongs to the σ-algebra generated by $X_1, X_2, \ldots, X_k$.

The counting process $\{N(t), t \geq 0\}$ is called a *renewal process.* Let the *renewal function* $m(t) = \mathsf{E}N(t)$.

THEOREM B.3.1 (The Elementary Renewal Theorem) *The expected average rate of renewals is defined by*

$$\lim_{t\to\infty} \frac{m(t)}{t} = \begin{cases} 1/\mu & \text{when } 0 < \mu < \infty, \\ 0 & \text{when } \mu = \infty. \end{cases}$$

B.4. Renewal Reward Processes

DEFINITION B.4.1 Suppose we are given a sequence of pairs of random variables (X_k, Z_k), where $X_k > 0$. Let $N(t)$ be the renewal process formed out of the sequence X_k as in Section B.3, that is, $N(t) = \max\{n : X_1 + X_2 + \ldots + X_n \leq t\}$. The process

$$Y(t) = \sum_{k=1}^{N(t)} Z_k \tag{B.1}$$

is called a *renewal reward process.*

In other words at the kth point of renewal, a *reward* of Z_k is added to the process. When $Z_k \equiv 1$ for each k, the renewal reward process coincides with the renewal process $N(t)$. The main theorem describing the limiting behavior of the renewal reward process is the following; (see Proposition 4.1 in Ross (1989)).

THEOREM B.4.1 *If* $Y(t)$ *is a renewal reward process given by* (B.1) *with* $\mathsf{E}(X), \mathsf{E}(Z) < \infty$, *where* X *and* Z *are generic random variables whose distributions are equal to those of* $\{X_k\}$ *and* $\{Z_k\}$, *respectively. Then,*

$$\lim_{t\to\infty} \frac{Y(t)}{t} = \lim_{t\to\infty} \frac{\mathsf{E}(Y(t))}{t} = \frac{\mathsf{E}Z}{\mathsf{E}X}.$$

B.5. Stochastic Dominance

DEFINITION B.5.1 Let ζ_1 and ζ_2 be two random variables with cumulative probability distributions Φ_1 and Φ_2, respectively. We say that ζ_2 *dominates* ζ_1, or equivalently, that ζ_2 is *stochastically greater than* ζ_1, if

$$\Phi_1(x) \geq \Phi_2(x) \qquad \forall x. \tag{B.2}$$

This concept of stochastic dominance was first introduced by Karlin (1958b) to compare stochastic demands. Heuristically, a demand ζ_2 is stochastically greater than a demand ζ_1, denoted as $\zeta_2 \underset{\text{st}}{\geq} \zeta_1$, if demands based on Φ_1 have a greater probability of taking smaller values than those based on Φ_2.

If $\zeta_2 \geq \zeta_1$, then ζ_2 is stochastically greater than ζ_1. The converse is true in the following sense. If ζ_2 is *stochastically greater than* ζ_1, then there exist a probability space $(\tilde{\Omega}, \tilde{\mathcal{F}}, \tilde{\mathsf{P}})$ and random variables $\tilde{\zeta}_1$ and $\tilde{\zeta}_2$ on this space having the same distribution functions Φ_1 and Φ_2, respectively, and satisfying $\tilde{\zeta}_2 \geq \tilde{\zeta}_1$. In fact, one can take $\tilde{\Omega}$ as an interval $[0, 1]$ with $\tilde{\mathcal{F}}$ being the set of all Borel sets and $\tilde{\mathsf{P}}$ being Lebesgue measure. For any $\omega \in \tilde{\Omega}$, we define $\tilde{\zeta}_i(\omega) = \Phi_i^{-1}(\omega), i = 1, 2$, where $\Phi^{-1}(y) = \inf\{x : \Phi(x) \geq y\}$ (if Φ is a continuous increasing function, then Φ^{-1} coincides with the usual inverse function). One can easily check that from $\Phi_1(x) \geq \Phi_2(x)$, follows $\Phi_1^{-1}(y) \leq \Phi_2^{-1}(y)$.

THEOREM B.5.1 *If ζ_2 dominates ζ_1, then*

$$\mathsf{E}\Psi(\zeta_1) \leq \mathsf{E}\Psi(\zeta_2) \tag{B.3}$$

for any nondecreasing real-valued function Ψ, for which both expectations exist.

Proof. See Veinott (1966). □

There are some other concepts of stochastic orderings between random variables. For example, the likelihood ratio ordering and the first moment ordering are also commonly used in the literature.

Let φ_1 and φ_2 denote the probability density functions of ζ_1 and ζ_2, respectively.

DEFINITION B.5.2 We say that ζ_2 is *greater than* ζ_1 in the sense of likelihood ratio ordering, if

$$\frac{\varphi_2(x)}{\varphi_1(x)} \leq \frac{\varphi_2(y)}{\varphi_1(y)}, \quad \forall x \leq y.$$

DEFINITION B.5.3 We say that ζ_2 is *greater than* ζ_1 in the sense of first moment ordering, if for a constant $C > 0$,

$$\varphi_2(x) = \varphi_1(x - C), \ \forall x \geq 0.$$

Clearly, these two types of stochastic ordering are stronger than the stochastic dominance ordering.

B.6. Markov Chains

Let $(\Omega, \mathcal{F}, \mathsf{P})$ be a probability space with a sequence of increasing σ-algebras $\mathcal{F}^0 \subseteq \mathcal{F}^1 \subseteq \ldots \subseteq \mathcal{F}^k \subseteq \mathcal{F}$ called *filtration*. Let $\mathbb{I}$ be a finite set $\{1, 2, \ldots, L\}$, called the *state space* and let $i_k \in \mathbb{I}, k = 0, 1, ...,$ be a sequence of random variables adapted to $\{\mathcal{F}^k\}_{k=0}^{\infty}$ (that is, i_k is $\mathcal{F}^k$-measurable). This sequence is called Markov chain with respect to the filtration $\{\mathcal{F}^k\}_{k=0}^{\infty}$, if

$$\mathsf{P}(i_{k+1} = j|\mathcal{F}^k) = \mathsf{P}(i_{k+1} = j|i_k), \quad k = 0, 1, ... \tag{B.4}$$

In addition, if

$$\mathsf{P}(i_{k+1} = j|\mathcal{F}^k) = \mathsf{P}(i_1 = j|i_0 = l)$$

on the set $\{i_k = l\}$, then the Markov chain is called *time-homogeneous.* When $\mathcal{F}^k$ is itself generated by the sequence $\{i_0, i_1, \ldots, i_k\}$, that is, when $\mathcal{F}^k = \sigma(i_0, i_1, \ldots, i_k)$, the definition of the Markov property takes the form

$$\begin{aligned}\mathsf{P}(i_{k+1} &= j|i_0 = l_0, i_1 = l_1, \ldots, i_{k-1} = l_{k-1}, i_k = l) \\ &= \mathsf{P}(i_{k+1} = j|i_k = l), \quad k = 0, 1, ...\end{aligned}$$

The quantities $p_{lj} = \mathsf{P}(i_1 = j|i_0 = l), \ l, j = 1, 2, \ldots, L$, are called *transition probabilities* and the matrix $P\{p_{lj}\}$ is called the *transition matrix* of the Markov chain $i_k, k = 1, 2 \ldots.$ The n-step transition probabilities $p_{lj}^{(n)}$ are defined as

$$p_{lj}^{(n)} = \mathsf{P}(i_{n+k} = j|i_k = l) = \mathsf{P}(i_n = j|i_0 = l).$$

The matrix $\{p_{lj}^{(n)}\}$ is equal to $\{p_{lj}\}^n$.

DEFINITION B.6.1 Two states j and l are said to be *communicating,* if there exist $n \geq 1$ and $m \geq 1$ such that $p_{lj}^{(n)} > 0$ and $p_{jl}^{(m)} > 0$. A Markov chain is *irreducible* if any two states j and l, $j, l = 1, 2 \ldots, L$, are communicating.

DEFINITION B.6.2 The period $d(j)$ of a state j is the greatest common divisor of the set $\{n : p_{jj}^{(n)} > 0\}$. If $d(j) = 1$, then the state j is *aperiodic.* Markov chain is said to have a period d, if all its states have the period d. If all states of the Markov chain are aperiodic, then the chain is said to be *aperiodic.*

The proof of the following theorem can be found in Section 2.7 of Hoel *et al.* (1972).

THEOREM B.6.1 *All states of an irreducible Markov chain have the same period.*

Denote by η_j the first hitting time of a state j, i.e.,

$$\eta_j = \min\{k : i_k = j\},$$

and let

$$m_{lj} = \mathsf{E}(\eta_j | i_0 = l).$$

DEFINITION B.6.3 A state j is *positive recurrent* if $m_{jj} < \infty$. A Markov chain is *positive recurrent* if all of its states are positive recurrent.

The proof of the next theorem is standard and it can be found in Section 2.4 of Hoel *et al.* (1972).

THEOREM B.6.2 *If a Markov chain is irreducible, then all of its states are positive recurrent. Moreover, for each two states j and l,*

$$m_{lj} < \infty.$$

DEFINITION B.6.4 A nonnegative sequence $\nu_1, \nu_2, \ldots, \nu_L$ is called a *stationary distribution* of a Markov chain, if $\sum_{l=1}^{L} \nu_l = 1$ and

$$\sum_{l=1}^{L} \nu_l \boldsymbol{p}_{lj} = \nu_j.$$

The proof of the following theorems can be found in Hoel *et al.* (1972), Sections 2.5 and 2.7.

THEOREM B.6.3 *For an irreducible finite state Markov chain, there exists a unique stationary distribution $\nu_1, \nu_2, \ldots, \nu_L$, with $\nu_l > 0, l = 1, 2 \ldots, L$.*

THEOREM B.6.4 *Let $i_0, i_1 \ldots$ be an irreducible Markov chain with period d. Let l and j be any two states in $\{1, 2, \ldots, L\}$ and a positive integer k be such that $p_{lj}^{(k)} > 0$. Then,*

$$\lim_{n\to\infty} p_{lj}^{(k+nd)} = d\nu_j,$$

where $\nu_1, \nu_2, \ldots, \nu_L$ is the unique stationary distribution for the Markov chain.

The conditional independence of the future and the past as it is expressed in (B.4) can also be applied in cases when n in (B.4) becomes a random variable. An important class of random variables for which the Markov property holds is the class of *stopping times.*

DEFINITION B.6.5 A random variable τ taking on nonnegative integer values is called a *stopping time* with respect to the filtration $\{\mathcal{F}^k\}$ if $\{\tau = k\} \subset \mathcal{F}^k$.

A σ-algebra $\mathcal{F}^\tau$ is the set of all events $A \in \mathcal{F}$ such that $A \cap \{\tau \leq k\} \in \mathcal{F}^k$.

The Markov property (B.4) remains valid if n in (B.4) is replaced by a stopping time τ; namely, the following theorem holds. Its proof can be found in Section II.9 of Chung (1967).

THEOREM B.6.5 *If τ is a stopping time, then*

$$\mathsf{P}(i_{\tau+1} = j|\mathcal{F}^\tau) = p_{lj} \quad on \quad \{i_\tau = l\}. \tag{B.5}$$

In this book we deal with the case when the filtration $\{\mathcal{F}^k\}$ is generated by a Markov chain $i_0, i_1, \ldots$ and a sequence of random variables $\xi_0, \xi_1, \ldots$ (that is, $\mathcal{F}^k = \sigma(i_0, i_1, \ldots, i_k; \xi_0, \xi_1, \ldots, \xi_{k-1})$), having the following properties. The random vector $(\xi_0, \xi_1, \ldots, \xi_k)$ and the sequence $(i_{k+1}, i_{k+2} \ldots)$ are independent. Conditioned on the event $\{i_0 = l_0, i_1 = l_1, \ldots, i_k = l_k\}$, the random variables $\xi_0, \xi_1, \ldots, \xi_k$ are independent, with the distribution functions $\Phi_{0,l_0}, \Phi_{1,l_1}, \ldots, \Phi_{k,l_k}$, respectively.

The proof of the following lemma is straightforward.

LEMMA B.6.1 *The sequence of random variables ξ_k, $k = 1, 2, \ldots$, is a Markov chain with respect to the filtration $\{\mathcal{F}^k\}$, where*

$$\mathcal{F}^k = \sigma(i_0, i_1 \ldots, i_k; \xi_0, \xi_1, \ldots, \xi_{k-1}).$$

Appendix C

CONVEX, QUASI-CONVEX AND K-CONVEX FUNCTIONS

C.1. PF_2 Density and Quasi-convex Functions

A brief description of the definition and some useful properties of PF_2 densities can be found in Porteus (1971) and Porteus (2002). We adopt a definition of a PF_2 density given by Ross (1983).

DEFINITION C.1.1 A density function $\phi : \mathbb{R} \to \mathbb{R}^+$ for which $\log \phi$ is concave is PF_2.

DEFINITION C.1.2 A function $f : \mathbb{R} \to \mathbb{R}$ is *quasi-convex* on a convex set $X \subset \mathbb{R}$ if for $x, y \in X$ and $0 \leq \lambda \leq 1$,

$$f(\lambda x + (1 - \lambda)y) \leq \max(f(x), f(y)).$$

It should be clear from the definition that the class of real-valued quasi-convex functions defined on $\mathbb{R}$ includes convex functions and monotone functions defined on $\mathbb{R}$.

Assume that a global minimum of a function f exists and let

$$f^* = \min_x f(x).$$

Let

$$m_f = \inf(x | f(x) = f^*).$$

It is clear that $f(x)$ is nonincreasing when $x < m_f$ and nondecreasing when $x \geq m_f$.

THEOREM C.1.1 *Let $f : \mathbb{R} \to \mathbb{R}$ be a quasi-convex function and let m_f be as defined above. Let ξ be a nonnegative random variable with a* PF_2 *density. Then,*

$$g(x) = \mathsf{E} f(x - \xi)$$

is a real-valued quasi-convex function defined on $\mathbb{R}$. Furthermore, there exists a minimizer of g at m_g such that $m_g \geq m_f$.

Proof. It follows from Lemma 9.6 in Porteus (2002) with $K = 0$. □

C.2. Convex and K-convex Functions

If g is a convex function, then

$$\lambda g(x) + (1-\lambda)g(u) \geq g(y) \tag{C.1}$$

for any $0 < \lambda < 1$, where $y = \lambda x + (1-\lambda)u$. From this definition we see that

$$(1-\lambda)(g(u) - g(x)) \geq g(y) - g(x).$$

Dividing both sides of this inequality by $(y-x)$ and observing that $y - x = (1-\lambda)(u-x)$, we get

$$\frac{g(u) - g(x)}{u - x} \geq \frac{g(y) - g(x)}{y - x}$$

for $x < y < u$. Thus, for a convex function, $\frac{g(u)-g(x)}{u-x}$ is an increasing function of u, and $\frac{g(u)-g(x)}{u-x}$ is a decreasing function of x.

THEOREM C.2.1 *Suppose $g(x)$ is a convex nonnegative function with linear growth, i.e. $0 \leq g(x) \leq C(1+|x|)$. Then, g is Lipschitz continuous on $(-\infty, \infty)$.*

Proof. For any $x > 0$ and any $y > x$,

$$\frac{g(y) - g(x)}{y - x} \leq \lim_{u\to\infty} \frac{g(u) - g(x)}{u - x} = \lim_{u\to\infty} \frac{g(u) - g(x)}{u} = \lim_{u\to\infty} \frac{g(u)}{u} \leq C.$$

The last inequality is a consequence of the linear growth condition. Similarly, by fixing y and letting $x \to -\infty$, we get

$$\frac{g(y) - g(x)}{y - x} \geq -C,$$

wherefrom Lipschitz continuity follows. □

Putting $x = y - b$ and $u = y + z$ (and as a result, $\lambda = z/(z+b)$) in (C.1), we have an equivalent formulation for convexity. For any y and any $b, z > 0$

$$\frac{b}{z+b} f(y+z) + \frac{z}{z+b} f(y-b) \geq g(y)$$

or

$$bg(z+y) + zg(y-b) \geq (z+b)g(y).$$

The property of K-convexity corresponds to the case when $g(z+y)$ in the LHS of the above inequality is replaced by $g(z+y) + K$ with $K \geq 0$. If we divide both sides of this inequality by b and then leave only $g(z+y) + K$ in the LHS, then we arrive at the following definition.

DEFINITION C.2.1 A function $g : \mathbb{R} \to \mathbb{R}$ is said to be *K-convex*, $K \geq 0$, if it satisfies the property

$$K + g(z+y) \geq g(y) + z\frac{g(y) - g(y-b)}{b}, \ \forall z \geq 0, b > 0, y. \qquad \text{(C.2)}$$

DEFINITION C.2.2 A function $g : \mathbb{R} \to \mathbb{D}$, where $\mathbb{D}$ is a convex subset $\mathbb{D} \subset \mathbb{R}$, is K-convex if (C.2) holds whenever $y + z, y$, and $y - b$ are in $\mathbb{D}$.

Required well-known results on K-convex functions or their extensions are collected in the following two theorems (see, e.g., Bertsekas (1976) or Bensoussan *et al.* (1983)).

THEOREM C.2.2

(a) *If $g : \mathbb{R} \to \mathbb{R}$ is K-convex, it is L-convex for any $L \geq K$. In particular, if g is convex, i.e., 0-convex, it is also K-convex for any $K \geq 0$.*

(b) *If g_1 is K-convex and g_2 is L-convex, then for $a, b \geq 0$, $ag_1 + bg_2$ is $(aK + bL)$-convex.*

(c) *If g is K-convex, and ξ is a random variable such that $\mathsf{E}|g(x-\xi)| < \infty$, then $\mathsf{E}g(x - \xi)$ is also K-convex.*

(d) *Restriction of g on any convex set $\mathbb{D} \subset \mathbb{R}$ is K-convex.*

(e) *Let $\{g_n\}_{n=1}^{\infty}$ be a sequence of K-convex functions converging pointwise to g. Then, g is also K-convex.*

Proof. Statements (a)-(c) are proved in Bertsekas (1976). Proofs of (d) and (e) are straightforward. □

THEOREM C.2.3 *Let $g : \mathbb{R} \to \mathbb{R}$ be a K-convex (l.s.c.) function such that $g(x) \to +\infty$ as $x \to +\infty$. Let A and B, $A \leq B$, be two extended real numbers with the understanding that the closed interval $[A, B]$ becomes open at A (or B) if $A = -\infty$ (or $B = \infty$). Let the notation $g(-\infty)$ denote the extended real number $\liminf_{x \to -\infty} g(x)$. Let*

$$g^* = \inf_{A \leq x \leq B} g(x) > -\infty. \qquad \text{(C.3)}$$

Let the function $h : (-\infty, B] \to \mathbb{R}$ be defined as

$$h(x) = \inf_{y \geq x, A \leq y \leq B} [K\mathbb{1}_{y>x} + g(y)].$$

Define the extended real numbers S and s, $S \geq s \geq -\infty$ as follows:

$$S = \min\{x \in \mathbb{R} \cup \{-\infty\} | g(x) = g^*, A \leq x \leq B\}, \qquad \text{(C.4)}$$

$$s = \min\{x \in \mathbb{R} \cup \{-\infty\} | g(x) \leq K + g(S), A \leq x \leq S\}. \tag{C.5}$$

Then,
(a) $g(x) \geq g(S)$, $\forall x \in [A, B]$;
(b) $g(x) \leq g(y) + K$, $\forall x, y$ *with* $s \leq x \leq y \leq B$;
(c) *the function*

$$h(x) \quad = \quad \left\{ \begin{array}{ll} K + g(S), & \text{for } x < s \\ g(x), & \text{for } s \leq x \leq B \end{array} \right\} \tag{C.6}$$

is l.s.c., and if g *is continuous,* $A = -\infty$, *and* $B = \infty$, *then* $h : \mathbb{R} \to \mathbb{R}$ *is continuous;*
(d) h *is* K*-convex on* $(-\infty, B]$.
Moreover, if $s > A$, *then*
(e) $g(s) = K + g(S)$;
(f) $g(x)$ *is strictly decreasing on* $(A, s]$.

Proof. When $s > A$, it is easy to see from (C.5) that (e) holds and that $g(x) > g(s)$ for $x \in (A, s)$. Furthermore, for $A < x_1 < x_2 < s$, K-convexity implies

$$K + g(S) \geq g(x_2) + \frac{S - x_2}{x_2 - x_1}[g(x_2) - g(x_1)].$$

According to (C.5), $g(x_2) > K + g(S)$ when $A < x_2 < s$. Then, it is easy to conclude that $g(x_2) < g(x_1)$. This proves (f).

As for (a), it follows directly from (C.3) and (C.4). Property (b) holds trivially for $x = y$, holds for $x = S$ in view of (a), and holds for $x = s$ since $g(s) \leq K + g(S) \leq K + g(y)$ from (C.5) and (a). We need now to examine two other cases: (1) $S < x < y \leq B$ and (2) $s < x < S, x < y \leq B$.

- Case 1 ($S < x < y \leq B$). Let $z = S$ if $S > -\infty$, and $z \in (S, x)$ if $S = -\infty$. By K-convexity of g, we have

$$K + g(y) \geq g(x) + \frac{y - x}{x - z}[g(x) - g(z)].$$

  Use (a) to conclude $K + g(y) \geq g(x)$ if $S > -\infty$. If $S = -\infty$, let $z \to -\infty$. Since $\liminf g(z) = g^* < \infty$, we can once again conclude $K + g(y) \geq g(x)$.

- Case 2 ($s < x < S, x < y \leq B$). If $s = -\infty$, let $z \in (-\infty, x)$. Then, $K + g(S) \geq g(x) + [(S - x)/(x - z)][g(x) - g(z)]$. Let $z \to -\infty$.

Since $\liminf g(z) \le K + g^* < \infty$, we can use (a) to conclude $g(x) \le K + g(S) \le K + g(y)$. If $s > -\infty$, then by K-convexity of g and (e),

$$K+g(S) \ge g(x)+\frac{S-x}{x-s}[g(x)-g(s)] \ge g(x)+\frac{S-x}{x-s}[g(x)-g(S)-K],$$

which leads to

$$\left(1+\frac{S-x}{x-s}\right)[K+g(S)] \ge \left(1+\frac{S-x}{x-s}\right)g(x).$$

Dividing both sides of the above inequality by $[1+(S-x)/(x-s)]$ and using (a), we obtain $g(x) \le K + g(S) \le K + g(y)$, which completes the proof of (b).

To prove (c), it follows easily from (a) and (b) that $h(x)$ equals the RHS. Moreover, since g is l.s.c. and $g(s) \le K + g(S)$, $h : \mathbb{R} \to \mathbb{R}$ is also l.s.c. The last part of (c) is obvious.

Next, if $s = -\infty$, then $h(x) = g(x)$, $x \in (-\infty, B]$, and (d) follows from Theorem C.2.2(d). Now, let $s > -\infty$. Therefore, $S > -\infty$. Note from (C.5) that $s \ge A$. To show (d), we need to verify

$$K + h(y+z) - \left[h(y) + z\frac{h(y)-h(y-b)}{b}\right] \ge 0$$

in the following four cases: (3) $s \le y-b < y \le y+z \le B$, (4) $y-b < y \le y+z < s$, (5) $y-b < s \le y \le y+z \le B$, and (6) $y-b < y < s \le y+z \le B$.

- Case 3 ($s \le y-b < y \le y+z \le B$) and Case 4 ($y-b < y \le y+z < s$). The proofs in these cases are obvious from the definition of $h(x)$.

- Case 5 ($y - b < s \le y \le y + z \le B$,). In this case, the definition of $h(x)$ and the fact that $g(s) \le g(S) + K$ imply

$$\begin{aligned} K + h(y+z) &- \left[h(y) + z\frac{h(y)-h(y-b)}{b}\right] \\ &= K + g(y+z) - \left[g(y) + z\frac{g(y)-g(S)-K}{b}\right] \\ &\ge K + g(y+z) - \left[g(y) + z\frac{g(y)-g(s)}{b}\right]. \qquad \text{(C.7)} \end{aligned}$$

  Clearly if $g(y) \le g(s)$, then the RHS of (C.7) is nonnegative in view of (b). If $g(y) > g(s)$, then $y > s$ and $b > y - s$, and we can use

K-convexity of g to conclude

$$\begin{aligned} K + g(y+z) &- \left[g(y) + z\frac{g(y)-g(s)}{b} \right] \\ &> K + g(y+z) - \left[g(y) + z\frac{g(y)-g(s)}{y-s} \right] \geq 0. \end{aligned}$$

- Case 6 ($y - b < y < s \leq y + z \leq B$). In this case, we have from the definition of $h(x)$ and (a) that

$$\begin{aligned} K + h(y+z) - \left[h(y) + z\frac{h(y)-h(y-b)}{b} \right] \\ = K + g(y+z) - [g(S) + K] \\ = g(y+z) - g(S) \geq 0. \end{aligned}$$

□

Remark C.2.1 Theorem C.2.3 is an important extension of similar results found in the literature; (see Lemma (d) in Bertsekas (1976), p.85). The extension allows us to easily prove the optimality of an (s, S) policy without imposing a condition like (2.11) for $x \to -\infty$ as discussed in Remark 2.1, and with the capacity constraints discussed in Section 2.8.2. Note that Theorem C.2.3 allows the possibility of $s = -\infty$ or $s = S = -\infty$; such an (s, S) pair simply means that it is optimal not to order.

Theorem C.2.4 *Let*

$$Q(x) = q(x) + g(x^+). \tag{C.8}$$

If $q(x)$ is convex and nonincreasing with $q(x) = 0$, $\forall x \geq 0$, $g(x)$ is K-convex, and $q^{'-}(0) \leq g^{'+}(0)$, where

$$g^{'+}(0) = \lim_{x \downarrow 0} \frac{\partial}{\partial x} g(x) \quad \textit{and} \quad q^{'-}(0) = \lim_{x \uparrow 0} \frac{\partial}{\partial x} q(x),$$

then $Q(x)$ is K-convex.

Proof. Clearly, we have $Q(x) = q(x) + g(0)$ for $x < 0$ and $Q(x) = g(x)$ for $x \geq 0$. We need to verify that

$$K + Q(y+z) - \left[Q(y) + z\frac{Q(y)-Q(y-b)}{b} \right] \geq 0, \quad \forall z \geq 0, b > 0, y, \tag{C.9}$$

in the following four cases: (1) $0 \leq y - b < y \leq y + z$, (2) $y - b < y \leq y + z < 0$, (3) $y - b < 0 \leq y \leq y + z$, and (4) $y - b < y < 0 \leq y + z$.

- Case 1 ($0 \leq y - b < y \leq y + z$) and Case 2 ($y - b < y \leq y + z < 0$). The proofs in these cases are obvious from the definition of $Q(x)$.

- Case 3 ($y - b < 0 \leq y \leq y + z$). In this case, we have from (C.8) that

$$K + Q(y + z) = K + g(y + z),$$

and

$$Q(y) + z\frac{Q(y) - Q(y - b)}{b} = g(y) + z\frac{g(y) - g(0) - q(y - b)}{b}.$$

Since $q(y - b) \geq 0$ and $y \leq b$, we have

$$\begin{aligned} K + g(y + z) &\geq g(y) + z\frac{g(y) - g(0)}{y} \\ &\geq g(y) + z\frac{g(y) - g(0) - q(y - b)}{b}. \end{aligned}$$

Therefore, (C.9) holds in case (3).

- Case 4 ($y - b < y < 0 \leq y + z$). In this case, the definition of $Q(x)$ implies that

$$K + Q(y + z) = K + g(y + z),$$

and

$$Q(y) + z\frac{Q(y) - Q(y - b)}{b} = q(y) + g(0) + z\frac{q(y) - q(y - b)}{b}.$$

From the K-convexity of $g(x)$, we know that for $0 \leq w \leq y + z$,

$$K + g(y + z) \geq g(w) + (y + z - w)\frac{g(w) - g(0)}{w}.$$

Letting $w \to 0$, we have

$$K + Q(y + z) \geq g(0) + (y + z)g^{'+}(0).$$

Since

$$g^{'+}(0) \geq q^{'-}(0) \geq q'(y) \geq \frac{q(y) - q(y - b)}{b},$$

and

$$yq'(y) \geq q(y), \forall y \leq 0,$$

we obtain

$$K + Q(y+z) \geq g(0) + (y+z)q'(y) \geq g(0) + q(y) + z\frac{q(y) - q(y-b)}{b}.$$

This validates (C.9) for Case 4. □

Gallego and Sethi (2005) generalize the concept of K-convexity to an n-dimensional Euclidean space. The resulting concept of $\mathcal{K}$-convexity, where $\mathcal{K} = (K_0, K_1, K_2, ..., K_n)$ with K_0 representing the joint setup cost and K_i representing the individual setup cost associated with product i, is useful in addressing production and inventory problems when there are individual product setup costs and/or joint setup costs.

References

ARAPOSTATHIS, A., V. S. BORKAR, E. FERNANDEZ-GAUCHERAND, M. K. GHOSH, AND S. I. MARCUS, Discrete Time Controlled Markov-Processes with Average Cost Criterion: A Survey, *SIAM Journal of Control and Optimization*, 31(2), 282-344, 1993.

ARROW, K. J., T. HARRIS, AND J. MARSCHAK, Optimal Inventory Policy, *Econometrica*, 29, 250-272, 1951.

ARROW, K. J., KARLIN, S., AND H. SCARF, *Studies in the Mathematical Theory of Inventory and Production*, Stanford, CA: Stanford University Press, 1958.

BALCER, Y., Partially Controlled Demand and Inventory Control: an Additive Model, *Naval Research Logistics Quarterly*, 30, 273-288, 1983.

BELLMAN, R., I. GLICKSBERG, AND O. GROSS, On the optimal inventory equation, *Management Science*, 2(1), 83-104, 1955.

BENSOUSSAN, A., M. CAKANYILDIRIM, J. A. MINJAREZ-SOSA, A. ROYAL, AND S. P. SETHI, Inventory Problems with Partially Observed Demands and Lost Sales, *Journal of Optimization Theory and Applications*, 136(3), 321-340, 2008a.

BENSOUSSAN, A., M. CAKANYILDIRIM, J. A. MINJAREZ-SOSA, S. P. SETHI, AND R, SHI, Partially Observed Inventory Systems: The Case of Rain Checks, *SIAM Journal of Control and Optimization*, 47(5), 2490-2519, 2008b.

BENSOUSSAN, A., M. CAKANYILDIRIM, A. ROYAL, AND S. P. SETHI, Bayesian and Adaptive Controls for a Newsvendor Facing Exponential Demand, *Journal on Decision and Risk Analysis*, 2009a.

BENSOUSSAN, A., M. CAKANYILDIRIM, AND S. P. SETHI, On the Optimal Control of Partially Observed Inventory Systems, *Comptes Rendus de l'Academié des Sciences Paris*, Ser. I 341, 419-426, 2005a.

BENSOUSSAN, A., M. CAKANYILDIRIM, AND S. P. SETHI, A Multiperiod Newsvendor Problem with Partially Observed Demand, *Mathematics of Operations Research*, 32(2), 322-344, 2007a.

BENSOUSSAN, A., M. CAKANYILDIRIM, AND S. P. SETHI, Optimal Ordering Policies for Inventory Problems with Dynamic Information Delays, *Production and Operations Management*, 16(2), 241-256, 2007b.

BENSOUSSAN, A., M. CAKANYILDIRIM, AND S. P. SETHI, Partially Observed Inventory Systems: The Case of Zero-Balance Walk, *SIAM Journal of Control and Optimization*, 46(1), 176-209, 2007c.

BENSOUSSAN, A., M. CAKANYILDIRIM, AND S. P. SETHI, A Note on 'The Censored Newsvendor and the Optimal Acquisition of Information,' *Operations Research*, 57(3), 2009b.

BENSOUSSAN, A., M. CROUHY, AND J. PROTH, *Mathematical Theory of Production Planning*, Amsterdam, The Netherlands: North-Holland, 1983.

BENSOUSSAN,

A., R. LIU, AND S. P. SETHI, Optimality of an (s, S) Policy with Compound Poisson and Diffusion Demands: A Quasi-variational Inequalities Approach, *SIAM Journal on Control and Optimization*, 44(5), 1650-1676, 2005b.

BERTSEKAS, D., *Dynamic Programming and Stochastic Control*, New York: Academic Press, 1976.

BERTSEKAS, D., AND S. E. SHREVE, *Stochastic Optimal Control: the Discrete Time Case*, New York: Academic Press, 1976.

BEYER, D., AND S. P. SETHI, Average Cost Optimality in Inventory Models with Markovian Demand, *Journal of Optimization Theory and Applications*, 92(3), 497-526, 1997.

BEYER, D., AND S. P. SETHI, The Classical Average-Cost Inventory Models of Iglehart and Veinott-Wagner Revisited, *Journal of Optimization Theory and Applications*, 101(3), 523-555, 1999.

BEYER, D., AND S. P. SETHI, Average-Cost Optimality in Inventory Models with Markovian Demands and Lost Sales, in *Analysis, Control and Optimization of Complex Dynamic Systems*, Edited by E.K. Boukas and R.P. Malhame, New York: Springer, 3-23, 2005.

BEYER, D., S. P. SETHI, AND R. SRIDHAR, Stochastic Multiproduct Inventory Models with Limited Storage, *Journal of Optimization Theory and Applications*, 111(3), 553-588, 2001.

BEYER, D., S. P. SETHI, AND M. TAKSAR, Inventory Models with Markovian Demands and Cost Functions of Polynomial Growth, *Journal of Optimization Theory and Application*, 98(2), 281-323, 1998.

CHEN, F., AND J. S. SONG, Optimal Policies for Multi-echelon Inventory Problems with Markov-modulated Demand, *Operations Research*, 49(2), 226-234, 2001.

CHEN, X., AND D. SIMCHI-LEVI, Coordinating Inventory Control and Pricing Strategies with Random Demand and Fixed Ordering Cost: The Infinite Horizon Case, *Mathematics of Operations Research*, 29(3), 698-723, 2004.

CHENG, F., Inventory Models with Markovian Demands, Doctoral Dissertation, University of Toronto, 1996.

CHENG, F., AND S. P. SETHI, A Periodic Review Inventory Model with Demand Influenced by Promotion Decisions, *Management Science*, 45(11), 1510-1523, 1999a.

CHENG, F., AND S. P. SETHI, Optimality of State-Dependent (s,S) Policies in Inventory Models with Markov-Modulated Demand and Lost Sales, *Production and Operations Management*, 8(2), 183-192, 1999b.

CHOW, Y. S., AND H. TEICHER, *Probability Theory*, New York: Springer-Verlag, 1978.

CHUNG, K. L., *Markov Chains with Stationary Transition Probabilities*, New York: Springer, 1967.

CHUNG, K. L., *A Course in Probability Theory*, New York: Academic Press, 1974.

DERMAN, C., Private Correspondence mentioned in Veinott and Wagner (1965), 1965.

DING, X., M. L. PUTERMAN, AND A. BISI, The Censored Newsvendor and the Optimal Acquisition of Information, *Operations Research*, 50, 517-527, 2002.

DVORETZKY, A., J. KIEFER, AND J. WOLFOWITZ, On the Optimal Character of the (s, S) Policy in Inventory Theory, *Econometrica*, 20, 586-596, 1953.

EDGEWORTH, F. Y., The Mathematical Theory of Banking, *Journal of Royal Statistical Society*, 51(1), 113-127, 1888.

ERLENKOTTER, D., An Early Classic Misplaced: Ford W. Harris' Economic Order Quantity Model of 1915, *Management Science*, 35(7), 898-900, 1989.

ERLENKOTTER, D., Ford Whitman Harris and the Economic Order Quantity Model, *Operations Research*, 38, 937-946, 1990.

FEDERGRUEN, A., AND P. ZIPKIN, An Efficient Algorithm for Computing Optimal (s, S) Policies, *Operations Research*, 32, 1268-1285, 1984.

FELLER, W. *An Introduction to Probability Theory and its Application*, Vol. 2, 2nd Edition, New York: Wiley, 1971.

FRIEDMAN, A., Foundation of Modern Analysis, New York: Holt, Rinehart and Wington, 1970.

FU, M., Sample Path Derivatives for (s, S) Inventory Systems, *Operations Research*, 42, 351-364, 1994.

GALLEGO, G., AND S. P. SETHI, K-Convexity in R^n, *Journal of Optimization Theory and Applications*, 127(1), 71-88, 2005.

GAVER, D. P., JR., On Base-stock Level Inventory Control, *Operations Research*, 7(6), 689-703, 1959.

GAVER, D. P., JR., Operating Characteristics of a Simple Production, Inventory-control Model, *Operations Research*, 9(5), 635-649, 1961.

HADLEY, G., AND T. WHITIN, *Analysis of Inventory Systems*, Englewood Cliffs, New Jersey: Prentice-Hall, 1963.

HARRIS, F., *Operations and Cost, (Factory Management Series)*, Chicago: A.W. Shaw and Company, 1913.

HERNÁNDEZ-LERMA, O., AND J. B. LASSERRE, *Discrete-Time Markov Control Processes*, New York: Springer, 1996.

HOEL, P. G., S. C. PORT, AND S.J. STONE, *Introduction to Stochastic Processes*, Waveland Press, 1972.

HOLT, C., F. MODIGLIANI, J. F. MUTH, AND H. A. SIMON, *Planning Production, Inventory, and Workforce*, Englewood Cliffs, New Jersey: Prentice-Hall, 1960.

HU, J. Q., S. NANANUKUL, AND W. B. GONG, A New Approach to (s, S) Inventory Systems, *Journal of Applied Probability*, 30, 898-912, 1993.

HUH, W. T., G. JANAKIRAMAN, AND M. NAGARAJAN, Average Cost Inventory Models: An Analysis Using a Vanishing Discount Approach, Working Paper, Columbia University, 2008.

IGLEHART, D., Optimality of (s, S) Policies in the Infinite Horizon Dynamic Inventory Problem, *Management Science*, 9(2), 259-267, 1963a.

IGLEHART, D., Dynamic Programming and Stationary Analysis of Inventory Problems, *Multistage Inventory Models and Techniques*, Edited by H. Scarf, D. Gilford, and M. Shelly, Stanford, CA: Stanford University Press, 1-31, 1963b.

IGLEHART, D., AND S. KARLIN, Optimal Policy for Dynamic Inventory Process Inventory Process With Nonstationary Stochastic Demands, *Studies in Applied Probability and Management Science*, Edited by K. Arrow, S. Karlin, and H. Scarf, Stanford, CA: Stanford University Press, 1962.

KALLENBERG, O., *Foundation of Modern Probability*, 2nd edition, New York: Springer, 2001.

KAPALKA, B. A., K. KATIRCIOGLU, AND M. L. PUTERMAN, Retail Inventory Control with Lost Sales, Service Constraints, and Fractional Lead Times, *Production and Operations Management*, 8(4), 1059-1478, 1999.

KARLIN, S., Steady State Solutions, *Studies in the Mathematical Theory of Inventory and Production*, Edited by K. Arrow, S. Karlin, and H. Scarf, Stanford, CA: Stanford University Press, 223-269, 1958a.

KARLIN, S., The Application of Renewal Theory to the Study of Inventory Policies, *Studies in the Mathematical Theory of Inventory and Production*, Edited by K. Arrow, S. Karlin, and H. Scarf, Stanford, CA: Stanford University Press, 270-297, 1958b.

KARLIN, S., Optimal Inventory Policy for the Arrow-Harris-Marshak Dynamic Model, *Studies in the Mathematical Theory of Inventory and Production*, Edited by K. Arrow, S. Karlin, and H. Scarf, Stanford, CA: Stanford University Press, 135-154, 1958c.

KARLIN, S., Dynamic Inventory Policy with Varying Stochastic Demands, *Management Science*, 6(3), 231-258, 1960.

KARLIN, S., AND C. CARR, Prices and Optimal Inventory Policies, *Studies in Applied Probability and Management Science*, Edited by K. Arrow, S. Karlin, and H. Scarf, Stanford, CA: Stanford University Press, 1962.

KARLIN, S., AND A. FABENS, The (s, S) Inventory Model under Markovian Demand Process, *Mathematical Methods in the Social Sciences*, Edited by K. Arrow, S. Karlin, and P. Suppes, Stanford, CA: Stanford University Press, 159-175, 1960.

KARLIN, S., AND H. SCARF, Inventory Models and Related Stochastic Processes. *Studies in the Mathematical Theory of Inventory and Production*, Edited by K. Arrow, S. Karlin, and H. Scarf, Stanford, CA: Stanford University Press, Chapter 17, 319-336, 1958.

KHOUJA, B., The Single Period (News-vendor) Problem: Literature Review and Suggestions for Future Research, *Omega*, 27, 537-553, 1999.

KRANTZ, S. A., *A Handbook of Real Variables*, New York: Birkhäuser, 2004.

KÜENLE, C. AND H. U. KÜENLE, Durchschnittskostenoptimale Strategien in Markovschen Entscheidungsmodellen bei unbeschränkten Kosten. *Mathematische Operationsforschung und Statistik, Series Optimization*, 8(4), 549-564, 1977.

LU, X., J. S. SONG, AND K. ZHU, On 'The Censored Newsvendor and the Optimal Acquisition of Information,' *Operations Research*, 53, 1024-1027, 2005.

MILLS, E., *Price, Output and Inventory Policy; A Study in the Economics of the Firm Industry*, New York: J. Wiley & Sons, 1962.

PARLAR, M., Y. WANG, AND Y. GERCHAK, A Periodic Review Inventory Model with Markovian Supply Availability, *International Journal of Production Economics*, 42(2), 131-136, 1995.

PORTEUS, E., On the Optimality of Generalized (s, S) Policies, *Management Science*, 17(7), 411-426, 1971.

PORTEUS, E., Numerical Comparison of Inventory Policies for Periodic Review Systems, *Operations Research*, 33, 134-152, 1985.

PORTEUS, E., *Foundations of Stochastic Inventory Theory*, Stanford, CA: Stanford University Press, 2002.

PRESMAN, E. AND S. P. SETHI, Inventory Models with Continuous and Poisson Demands and Discounted and Average Costs, *Production and Operations Management*, 15(2), 279-293, 2006.

RAMAN, A., N. DEHORATIUS, AND Z. TON, Execution: The Missing Link in Retail Operations, *California Management Review*, 43(3), 136-152, 2001.

RAYMOND, F. E., *Quantity and Economy in Manufacture*, New York: McGraw-Hill, 1931.

ROCKAFELLAR, R. T., *Convex Analysis*, Princeton, NJ: Princeton University Press, 1970.

ROSS, K. A., *Elementary Analysis: The Theory of Calculus,* New York: Springer, 1980.

ROSS, S. M., *Stochastic Processes,* Wiley Series in Probability and Statistics, New York: Wiley, 1983.

ROSS, S. M., *Introduction to Probability Models*, San Diego, CA: Academic Press, 1989.

RUDIN, W., *Principles of Mathematical Analysis*, New York: McGraw-Hill, 1976.

SAHIN, I., *Regeneration Inventory Systems: Operating Characteristics and Optimization*, Bilkent University Lecture Series, New York: Springer-Verlag, 1990.

SCARF, H., The Optimality of (s, S) Policies in the Dynamic Inventory Problem, *Mathematical Methods in the Social Sciences*, Edited by K. Arrow, S. Karlin, and P. Suppes, Stanford, CA: Stanford University Press, 196-202, 1960.

SCHÄL, M., On the Optimality of (s, S) Policies in Dynamic Inventory Models with Finite Horizon, *SIAM Journal on Applied Mathematics*, 30(3), 528-537, 1976.

SETHI, S. P., AND F. CHENG, Optimality of (s, S) Policies in Inventory Models with Markovian Demand, *Operations Research*, 45(6), 931-939, 1997.

SETHI, S. P., H. YAN, AND H. ZHANG, Inventory Models with Fixed Costs, Forecast Updates, and Two Delivery Modes, *Operations Research*, 51(2), 321-328, 2003.

SETHI, S. P., H. YAN, AND H. ZHANG, *Inventory and Supply Chain Decisions with Forecast Updates*, New York: Springer, 2005a.

SETHI, S. P., AND Q. ZHANG, *Hierarchical Decision Making in Stochastic Manufacturing Systems*, Boston, MA: Birkhäuser, 1994.

SETHI, S. P., AND Q. ZHANG, Multilevel Hierarchical Decision Making in Stochastic Marketing-Production Systems, *SIAM Journal on Control and Optimization*, 33(2), 528-553, 1995.

SETHI, S. P., H. ZHANG, AND Q. ZHANG, *Average-Cost Control of Stochastic Manufacturing Systems*, New York, NY: Springer, 2005b.

SHAKED, M., AND J. G. SHANTHIKUMAR, *Stochastic Order and Their Applications*, New York: Academic Press, 1984.

SHREVE, S. E., Abbreviated proof [in the lost sales case], *Dynamic Programming and Stochastic Control,* Edited by D.P. Bertsekas, New York: Academic Press, 105-106, 1976.

SILVER, E., Operations Research in Inventory Management: a Review and Critique, *Operations Research*, 29, 628-645, 1981.

SOGOMONIAN, A., AND C. TANG, A Modeling Framework for Coordinating Promotion and Production Decisions within a Firm, *Management Science*, 39(2), 191-203, 1993.

SONG, J. S., AND P. ZIPKIN, Inventory Control in a Fluctuating Demand Environment, *Operations Research*, 41, 351-370, 1993.

STIDHAM, S. JR., Cost Models for Stochastic Clearing Systems, *Operations Research*, 25(1), 100-127, 1977.

SZNAJDER, R., AND J. A. FILAR, Some Comments on a Theorem of Hardy and Littlewood, *Journal of Optimization Theory and Applications*, 75, 201-208, 1992.

THOWSEN, G. T., A Dynamic Nonstationary Inventory Problem for a Price/Quantity Setting Firm, *Naval Research Logistics Quarterly*, 22, 461-476, 1975.

TIJMS, H., Analysis of (s, S) inventory models, *Mathematical Centre Tracts*, 40, Amsterdam, The Netherlands: Mathematisch Centrum, 1972.

VEINOTT, A., The Optimal Inventory Policy for Batch Ordering, *Operations Research*, 13, 1103-1145, 1965.

VEINOTT, A., On the Optimality of (s, S) Inventory Policies: New Conditions and a New Proof, *SIAM Journal on Applied Mathematics*, 14(5), 1067-1083, 1966.

VEINOTT, A., AND H. WAGNER, Computing Optimal (s, S) policies, *Management Science*, 11, 525-552, 1965.

WAGNER, H., Comments on "Dynamic Version of the Economic Lot Size Model," *Management Science*, 50(12) Supplement, 1775-1777, December 2004.

WAGNER, H., AND T. WHITIN, Dynamic Version of the Economic Lot Size Model, *Management Science*, 5, 89-96, 1958; republished in *Management Science*, 50(12) Supplement, 1770-1774, December 2004.

WIJNGAARD, J., *Stationary Markovian Decision Problems.* Doctoral Theses, Technische Hogeschool Eindhoven, The Netherlands, 1975.

YOSIDA, K., *Functional Analysis*, 6th Edition, New York: Springer-Verlag, 1980.

YOUNG, L., Price, Inventory and the Structure of Uncertain Demand, *New Zealand Journal of Operations Research*, 6, 157-177, 1978.

ZABEL, E., Monopoly and Uncertainty, *The Review of Economic Studies*, 37, 205-219, 1970.

ZHENG, Y. S., A Simple Proof for Optimality of (s, S) Policies in Infinite-Horizon Inventory Systems, *Journal of Applied Probability*, 28, 802-810, 1991.

ZHENG, Y. S., AND A. FEDERGRUEN, Finding Optimal (s, S) Policies Is About As Simple as Evaluating a Single Policy, *Operations Research*, 39(4), 654-665, 1991.

ZIPKIN, P., *Foundations of Inventory Management*, New York: McGraw-Hill, 2000.

ZIPKIN, P., On the Structure of Lost-Sales Inventory Model, *Operations Research*, 56(4), 937-944, 2008a.

ZIPKIN, P., Old and New Methods for Lost-Sales Inventory Systems, *Operations Research*, 56(5), 1256-1263, 2008b.

Copyright Permissions

Selected portions of the publications below have been reprinted with permissions as indicated.

Author Index

Subject Index

MIX
Papier aus verantwortungsvollen Quellen
Paper from responsible sources
FSC® C105338

If you have any concerns about our products,
you can contact us on
ProductSafety@springernature.com

In case Publisher is established outside the EU,
the EU authorized representative is:
Springer Nature Customer Service Center GmbH
Europaplatz 3, 69115 Heidelberg, Germany

Printed by Libri Plureos GmbH
in Hamburg, Germany